Remnants
of the Fall
Revelations of
Particle Secrets

Remnants
of the Fall
Revelations of
Particle Secrets

William B Rolnick

Wayne State University
USA

World Scientific
New Jersey • London • Singapore • Hong Kong

Published by

World Scientific Publishing Co. Pte. Ltd.

5 Toh Tuck Link, Singapore 596224

USA office: Suite 202, 1060 Main Street, River Edge, NJ 07661

UK office: 57 Shelton Street, Covent Garden, London WC2H 9HE

British Library Cataloguing-in-Publication Data
A catalogue record for this book is available from the British Library.

REMNANTS OF THE FALL: REVELATIONS OF PARTICLE SECRETS

ISBN 981-238-060-4
ISBN 981-238-146-5 (pbk)

Printed by FuIsland Offset Printing (S) Pte Ltd, Singapore

This book is dedicated to my deceased son Stefan who is missed with heart-rending sorrow.

CONTENTS

THE POEMS

OVERTURE

The Descent of Symmetry — for narrator and chorus

PART I THE JOURNEY BEGINS

PART II ENTERING SYMMETRY LAND

*This limerick was a prize-winner in the contest held by the American Physical Society, and appeared in the March '97 issue of the APS NEWS.

PART III ON TO THE STANDARD MODEL AND BEYOND

FINALE

SYMMETRY blink GRAVITY

DEAR READER

The purpose of this book is to present our understanding of the basic building blocks of nature and their fundamental interactions. The underlying ideas have a beauty and power worthy of the attention of all those who are curious about the world. These concepts, along with the experimental investigations associated with them, are among the greatest achievements of human thought. Although they are usually written in the language of mathematics, I have presented their essences without its use. This should make the ideas accessible to interested readers whose backgrounds may not include advanced mathematics.

In this book, the reader is taken along the path of knowledge that leads from our basic physical concepts to our understanding of the fundamental interactions of the underlying building blocks of nature. That understanding was developed in the second half of the twentieth century and is incorporated in the theory known as the Standard Model, which has been verified in many different high-precision experiments. The magnificent power of the particle-type symmetries that surreptitiously govern the fundamental interactions will be revealed here. Thereafter, the reader will be introduced to the dreams that contain even broader, more fundamental conjectures. Along the way, we will be exploring the fundamental building blocks of matter (quarks and leptons), and the concepts of relativity, quantum theory, symmetry, gauge theories (and their associated particles), the Standard Model (theory) and even beyond — the theoretical ideas and their experimental foundations and verifications. The last part of the book provides a brief introduction to the experimental enterprise of particle (otherwise known as "high-energy") physicists, and the connections of experimental studies to the theories we have met on our journey. The reader should feel free to read that part whenever curiosity demands, without feeling that all of the first three parts must be read first.

If you haven't traveled this way before, you are entering a new country with an unfamiliar culture, so you may not understand everything you encounter. Nevertheless, I hope you find your trip intriguing. If so, you may even enjoy retracing your steps or following other paths to these fundamental concepts. Note that there is a Glossary at the end of the book to help you along the way, by reminding you of those concepts and definitions already read.

Some readers of a *preliminary* version of this book, from the arts, literature, humanities and business world, found the initial flow of unfamiliar words somewhat

disconcerting. I ask those readers who have a similar initial reaction to bear with me; all the important words will be repeated along with reminders of their meanings. Furthermore the Glossary should be useful in this regard.

If you listen to National Public Radio or watch Jim Lehrer's News Hour, you should recall that a short musical interlude is used at the end of each story. I have adapted this technique by inserting poems after intense discussions. These verses serve several useful purposes. They provide some respite and hopefully some amusement, as well as reinforce your grasp of the concepts just explained.

The book begins with a poem that sets the stage for all that follows, much like an overture to an opera. It dramatizes the "grand unified" origin of the particles of our universe, believed to have occurred very shortly after its emergence from the Big Bang. The particular names of the particles involved can be accepted just as we would accept unfamiliar names of characters in a play. By the end of the book, you will be familiar with them and their corresponding properties. That is *my* obligation to you and not *vice versa*.

Our knowledge of the thought of people in ancient times has been culled from their writings, as well as their oral traditions. In those contemplations, we see a mixture of philosophy, science, history, and theology. During the Renaissance, scientific contemplation separated itself from the others. Its scope was restricted to those concepts that can be tested by directly measurable observations of nature. These included both passive observation and experimental investigations. (This restrictive demand upon scientific inquiry was succinctly expressed in the writings of Francis Bacon.*) Since that time, so many centuries ago, a widening gulf has developed between the realm of scientific thought and the so-called "humanities." (The existence and width of that gulf was brought to public consciousness by the incisive lectures of C.P. Snow, which are available in book form.) There have been several very fine books and articles written, as well as public lectures given, to acquaint those not conversant with the concepts of science with their power and beauty. It is my hope that this book will serve as another span across that gulf between the "two cultures," making the magnificent concepts that have been developed in our pursuit of the *underlying* structure and workings of nature accessible to those who do not have the time or inclination to master the mathematics involved.

I have not named the creators of the various concepts described here, because there are many people who deserve credit and the historical record is not the subject of this book. I do hold those brilliant, insightful people in high regard and am truly amazed at the beautiful structure of nature that their fertile imaginations have uncovered. (The poems that humorously portray Sherlock Holmes' and Hercule Poirot's imagined involvement should not be seen as in any way detracting from the true creators of the ideas discussed.) I strongly recommend the book "Inward Bound" by A. Pais (as well as his earlier books about Einstein, etc.) as an

*in *Novum Organum*; the relevant sections are reprinted in *The World of Physics*, Simon & Schuster (1987).

excellent, accurate and beautiful historical account of the development of the ideas of particle physics during the first eighty odd years of the twentieth century. "The Second Creation" by R.P. Crease and C.C. Mann and "Constructing Quarks" by A. Pickering include the developments in the early 1990s. There are several more recent books that cover the developments in the 1990s. I have led the reader to the threshold of the unifying theory of superstrings, and recommend the celebrated book that explains those ideas and their incorporation of gravitation into a *fully* unified theory: "The Elegant Universe" by Brian Greene. The books I have cited here have many excellent references as well.

It is my pleasure to acknowledge the people, from many different walks of life, who read parts of various drafts of this book and gave me both their reactions and their support. A hearty thank you goes to Lynne Hirschelman, Ray Greene, Al Varone, J. Will Campbell, Rita Winters, Francine Nightingale, Trevor Nightingale as well as my wife Ellen. I also wish to thank my fellow physicists for their advice. Thanks goes to Alexey Petrov, Dave Cinabro, J.J. Chang, Jim M. Johnson, and Bert Stein.

INTRODUCTION

In this book, we will retrace the age-old, intriguing quest, as pursued by scientists in the 20$^{\text{th}}$ and on into the 21$^{\text{st}}$ century, for the *essence* of nature — the fundamental building blocks and their interactions. We will also glimpse beyond the theories that have already been experimentally verified, at the major conjectures that have been proposed but not tested or fully developed.

We will first describe the background necessary for the comprehension of our present understanding, and then will proceed to explore the beautiful and powerful symmetry theories that account for nature's underlying interactions. We will see that symmetries of particle type are analogous to the beautiful spatial arrangements of symmetric objects and consequently possess an arcane beauty of their own. We will see how experimental clues and theoretical insights, in a symbiotic dance, have led to the discovery of those hidden, now broken, symmetries that we believe prevailed when our universe was very young and very hot. We will *finally* explore the surmise, which results from our current understanding, that the types of particles that make up the universe were not the first kinds of matter that existed. Indeed, for an instant just after the Big Bang, over-arching particle-type symmetries should have held sway over the fundamental particles that roamed freely between their interactions with one another. We believe that as the universe rapidly expanded and cooled, those symmetries were broken in very special ways that left the composite strongly interacting particles inhabiting our universe today and the presently observed interactions as *remnants*. And consequently, you & I and all we "see" are remnants of that fall.

OVERTURE

THE DESCENT OF SYMMETRY

for narrator and chorus

<u>Program Notes</u>

- The characters in this overture are particles, but I ask you to think of them as you would the characters of *any* play. Here, their properties should be thought of as parts of their respective IDs.

- The *grand unified symmetry* in the opening of this drama is a symmetry of particle behavior — all matter particles had interactions that *appeared* identical. This symmetry of behavior resulted from the *masslessness* of all the particles of the interactions and *their* equivalence to one another. (The matter particles were massless as well.)

- The property called "color" has nothing to do with actual color; it is, instead, the nuclear-force "charge" analogous to electric charge.

- The spin property results from the fact that the particles *behave* as if they are rotating, much like the Earth rotates about its axis.

<u>Cast of Characters</u>*

The fermions - the fundamental-matter (spin-1/2) guys

quarks (vowels pronounced like "quart") fermions of "color" (associated with nuclear forces)
leptons colorless fermions (like the electron)

The gaugons (rhymes with "rage on") - particles of the interaction that shmooze with fermions and sometimes with one another (<u>spin-1 except for the graviton</u>)

gluons "colorful" gaugons of nuclear forces
photon a gaugon of electromagnetism, including light
W charged gaugons of radioactivity
Z an uncharged gaugon of radioactivity
gravitons (spin-2) gaugons of gravity

The hadrons - quark clumps

mesons (the 'e' is pronounced as in "met") shy types of hadrons (discovered in the 1940s)
baryons the proud and famous proton and neutron, and their heavier cousins

Others

Higgses trouble-making strangers
gluonium a character made of two lovingly-entwined *gluons*

*Note to the cognoscenti: Supersymmetric partners of all the characters here may be lurking in the shadows; they do not take part in the action of our drama.

fermions - here, the fundamental particles of *matter: quarks* (of color) and **leptons** (like the electron)
spin-1 gaugons - gluons (of color), **photons, Ws, Zs**, that carry the interactions between matter particles.
hadrons - bound states of *quarks* and/or *antiquarks* **gluonium** - two-*gluon* bound state **graviton** - gaugon
of gravity.

Before the ancient times, ago
before the time before,
just after time began to go
and space was little more,

the universe had symmetry
both local and so grand —
though there was no geography;
there was no sea or land.

chorus: Oh symmetry, oh symmetry,
your beauty does astound,
and all of us are filled with glee
whenever you are found.

'Twas hotter then than hot can be
'twas hotter than the sun,
and all there was, was *energy;*
the "world" had just begun.

And all was ruled by symmetry
unbroken and supreme,
and "particles" began to *be*
as in a fev'rish dream.

Some say they're strings 'va grand-viol
that nature plucks and plays.
Each different tone has its own role,
as particle 't portrays.

And so the particles we know
are superstrings, perchance.
But it was oh so long ago
that they could twist and dance.

For stiffen did those superstrings,
in universe, out-bound,
as quarks, as leptons wandering
or gaugons rushing 'round.

chorus: For all space just keeps expanding
and the temp'rature does fall.
The inev'table demanding
of the laws that govern all.

And so it was long, *long* ago,
the universe aboil,
that superstrings had to forgo
their craving to uncoil.

Grand symmetry, so mighty then,
its dominance proclaimed.
And all the particles we ken
were caught in its grand frames.

chorus: Oh symmetry, oh symmetry,
your beauty does astound,
and all of us are filled with glee
whenever you are found.

The fermions they had no mass
and spin one-half had they.
But diff'rences 'tween them, alas,
would soon come into play.

And many gaugons there arose
from th'interaction's play.
So many different types of those
but much alike were they.

All spin-one gaugons, full of flair,
did crave for fermions.
And some did even make a pair
of those desired ones.

These spin-one ones in *virtual* guise
'tween fermions did go.
And carried interactions' ties
from one to th'other so.

And thus they functioned in their course
as interactions' hands.
Their *masslessness,* it was the source
of th' symmetry, so grand.

chorus: Oh symmetry, oh symmetry,
your beauty does astound,
and all of us are filled with glee
whenever you are found.

As space expanded full and all,
the particles were stirred.
The temp'rature did fall and fall
and then a *change* occurred.

We're told that hidden strangers struck:
the mighty Higgses bold.
From vacuum, some mass they plucked
for many to enfold.

continued over

fermions - here, the fundamental particles of *matter:* **quarks** (of color) and **leptons** (like the electron)
spin-1 gaugons - **gluons** (of color), **photons**, **Ws**, **Zs**, that carry the interactions between matter particles.
hadrons - bound states of *quarks* and/or *antiquarks* **gluonium** - two-*gluon* bound state **graviton** - gaugon
of gravity.

The change was sudden and extreme:
　some gaugons now had mass.
Yes, just as in a horr'ble dream
　this furor came to pass.

The symmetry was broken so,
fragmented through and through.
But part remained to govern, though
much more was lost, 'tis true.

Yes, symmetry survived the break
　but something wasn't right!
For then its parts, they did forsake
　equality of might.

chorus: And all space just keeps expanding
and the temp'rature does fall.
The inev'table demanding
of the laws that govern all.

Now, for an instant, there remained
　a symmetry quite wide.
"Electroweak," that was its name,
　but cooling broke its pride.

Small symmetries *still* seemed to be;
of diff'rent strengths were they.
　But fermions' equality
　this world did not obey.

And then the quarks did so denounce
　the leptons *weak* reply.
Their *strong* outlook made them renounce
　the leptons, by and by.

But soon another time would come
　with great catastrophe.
The quarks would lose all their freedom
　while leptons remained free.

Beware those gluons, they were told.
　Beware their strength and ken.
For once they get you in their fold
　You'll not get out again.

chorus: And all space just keeps expanding
and the temp'rature does fall.
The inev'table demanding
of the laws that govern all.

And bags were formed of gluon fields;
　three quarks locked in each one.
The forces that each gluon wields
　thus made the *baryon.*

And when a quark met antiquark,
　a *meson* (bound) to make,
these fermions tried hard to balk.
　Their freedom was at stake.

But finally all quarks were bound
　and antiquarks as well.
And gluons were together wound,
　their colors thus to quell.

So color passed out of our world
　with no apparent trace.
Gluonium and hadrons swirled.
　No color showed its face.

But hidden deep inside these lumps,
　within them there's no doubt,
the color force rages and romps
　and some effect leaps out.

'Tis stronger though than might be thought,
　the strongest force — deluxe.
Yet just a remnant of that caught
　in each bag's web of flux.

chorus: And all space just keeps expanding
and the temp'rature does fall.
The inev'table demanding
of the laws that govern all.

And many gaugon types were gone,
　too heavy to complain.
But photons still do carry on,
　for massless they remain.

To bring 'lectromagnetic force
　'tween particles with charge,
a photon loves to fly the course
　o'er lengths, from small to large.

And how did all the leptons fare?
　Did they survive as well?
By shunning gluons with great care,
　they're free, though th' others fell.

continued over

fermions - here, the fundamental particles of *matter:* **quarks** (of color) and **leptons** (like the electron)
spin-1 gaugons - *gluons* (of color), **photons, Ws, Zs**, that carry the interactions between matter particles.
hadrons - bound states of *quarks* and/or *antiquarks* **gluonium** - two-*gluon* bound state **graviton** - gaugon
of gravity.

But charg'd leptons are not *all* free.
To interact 's their plight,
with photons that ignore their plea
sometimes, and bind them tight.

A weak force did survive, also,
the symmetry's descent.
Its gaugons' masses rather low;
its power not all spent.

The W 'n Z, those gaugons weak,
with leptons interact.
And in the lumps, to quarks they streak
their force-role to enact.

(These weak-force gaugons look at spin
and mingle with great care.
With lefties only will they grin,
with righties they don't dare.)

And many weak decays, arranged
by Ws, occurred.
The flavors of the quarks were changed,
then they were reinterred.

Thus many hadrons did decay,
so many types to few.
The baryons' large type-array
became now only two.

The proton and the neutron, "white",
were all the kinds there left.
No quarks or gluons were in sight
Of color 'twas bereft.

And mesons, "white", were there anon,
with colors in their glue.
But they decayed, all color gone,
to lepton-photon stew.

And fourth among the forces left
without depravity,
with gravitons that link all heft,
pervaded gravity.

And now our story does begin,
we, remnants of the fall:
THE BEGINNING*

*This is *the beginning* spoken of in the first book of the Bible: B'RAYSHIYT (meaning "In The
Beginning"), translated as *Genesis.*

Part I

THE JOURNEY BEGINS

Chapter 1

ABOUT QUESTIONS AND ANSWERS

Even in prehistoric times, our ancestors wondered about the world in which they existed — what is it made of and how does it work? These kinds of questions are still being considered today in such diverse fields as philosophy, theology, poetry, drama, literature and science. Join me on a journey to the smallest parts of the *stuff* of the world, accessible to modern science, as we seek the structure and the fundamental interactions of the world around us — as we probe, in the words of Francis Bacon,* "the very entrails of nature." During this trip, we will see that knowledge of the very small actually helps us understand the behavior of the universe at very early times. As we traverse the paths prepared for us, we will find this age-old quest to be part mystery story and part adventure.

Many researchers over many centuries have cut their way through the unknown to bring us to our present level of understanding. Let us first explore the early part of the modern path by asking: what is scientific knowledge? There are several ways in which people claim to "know" things. You might intuitively know, or know deep in your heart, or feel it with every fiber of your being. However, others might have serious doubts about such "knowledge." In contrast, *scientific* knowledge is knowledge based on *repeatable, verifiable* experimental tests. This kind of knowledge is constantly being tested and challenged by new experiments. In fact, it would delight any scientist to discover something new or even to show that a cherished concept or result is wrong! Becoming a major player in the search for an understanding of the workings of nature, along with the excitement, the fame and the prestige that would ensue would more than "make their day!" Consequently there is no "scripture" in modern science; every concept is open to question. Indeed, the scientific "revolutions" of the past (those associated with Galileo, Newton, Einstein, Bohr, etc.) showed the limitations of previous concepts and rested firmly on precise observations. Nature itself was the final arbiter.

*in his *Novum Organum*, an early seventeenth century work reprinted in *The World of Physics* by Jefferson Hane Weaver, Simon and Schuster (1987).

One of the age-old profound questions amenable to scientific inquiry is: What is the physical matter of the world made of and what is holding those constituents together? If we cut a chunk of matter into smaller and smaller snips, do we always get pieces of the same substance or do we finally reach a smallest piece? If there is a smallest piece, can it be broken into more elementary constituents? — What are little snips made of?* — Is there a fundamental set of "building blocks" from which everything is made? And, if so, how do they *interact* with one another?

In this book we will explore the modern version of these contemplations on the structure of nature, namely the science of elementary particle physics: What is a science? — What fundamental questions does elementary particle physics address? — How are these questions asked and how are they answered? We will see that there is some philosophy, some poetry and indeed some drama present in the enterprise.

First a few words about modern science. In modern science, theories about the structure and workings of nature are proposed, which often have a philosophy and poetry of their own. The language used is mathematics (which has its own arcane beauty), because of its great precision and extraordinary power of expression. The logical structure of mathematical equations implies relationships that are necessarily true. This *strict connection* between the phenomena associated with various mathematical expressions is the strength and the beauty of mathematical formulations of theories. (A set of equations that merely *mimic* observations, treating them as independent facts, would never qualify as a theory!) Once such a theory is formulated, there is little that can be modified within it, although the discovery of the ramifications and predictions of the theory is an engrossing task. But to ascertain whether the theories are *valid* requires confrontation with nature itself. This "confrontation" manifests itself in the form of experiments. In order to earn its credentials as a *scientific* theory, a proposed theory must lead to testable predictions about the *world*. These predictions must then be implemented as questions to be answered by experiment, questions whose answers can agree or *disagree* with the predictions. In other words, both confirming and violating results must be possible. A newly proposed theory must be consistent with all verified observations as well as predict and explain new phenomena. This is often achieved by having the theory mathematically reduce in appropriate situations to older theories, which explained all of the previously ascertained facts about those situations. The human imagination is very creative and once the mathematics used in science is mastered, many possible theoretical structures can be thought of. However, unless we can coax predictions from them that lead to feasible experimental tests, their new features are not of a scientific nature — they fall into the realm of *metaphysics*, which is *not on our itinerary*.

If there are two different theories that produce exactly the same predictions for *all* their experimental tests, we might wish to seek other criteria to determine

*A nursery rhyme asks and answers: "What are little boys made of?... Snips and snails and puppy dog tails..." and so we ask, "What are little *snips* made of?"

which theory to adopt. Let us examine two examples from the past, to gain some insight into the problem and how it is handled. The first example is a comparison of the theory (embraced by theologians in the Middle Ages) that the earth is the center about which all heavenly bodies move, to Copernicus' sun-centered solar system. By the use of imaginary circles (called epicycles) with centers moving around larger imaginary circles, the motions of the planets could be matched by calculations in the Earth-centered theory. However, almost every new discovery or measurement required the incorporation of more epicycles, so that the final theory appeared highly contrived. It made no *new* testable predictions and was little more than a complicated way of describing observations. On the other hand, Copernicus' sun-centered solar system explained the observations, including the apparent backward motion in the sky sometimes exhibited by planets, as a *natural consequence* of the theory. Furthermore, it led to the understanding of force, motion, and gravity that was later developed by Newton into a full mathematical theory. In fact, Newton's theory could explain the motion of objects on Earth as well as that of the "heavenly" bodies — achieving a synthesis of two apparently disparate fields of study. Thus, phenomena, which appeared to be completely different, were shown to be aspects of the same underlying laws of nature. It is interesting to note that the original theory of Copernicus could not match detailed experimental observations as well as the theory of epicycles, but Newton's elegant mathematical formulation of gravitation (less than a century later) matched the experimental observations beautifully. Newton's theory had a few underlying basic concepts, from which very many predictions flow, as well as a general framework that serves as the paradigm of modern science. It even had a profound effect on philosophy, giving rise to the mechanistic, deterministic world view.

The second example is a comparison of two different formulations of quantum theory, originally created in the 1920s to explain the behavior of atoms and other phenomena at distances comparable to atomic sizes. Both are formulated along the lines established by Newton, namely basic mathematical relationships that can explain many different types of situations and experiments. These two quantum theories appeared quite different. (One was formulated using very advanced algebra called matrix algebra, the other using the advanced calculus of differential equations.) For every experimental observation, the calculation with each theory gave *exactly* the same result. The dilemma as to which theory should be adopted was resolved, when it was shown that they are mathematically equivalent; one could be mathematically derived from the other. Each is a different formulation of the same theory called *quantum mechanics*.

By now, you have probably realized that we are operating under the unspoken assumption that there is a logical consistency to nature, for without it we have no science at all. We would have a mysticism appropriate to theology instead, which is not part of our itinerary.

Now let us take a few moments to kick around the question of the connection between our physical theories and REALITY. If we construct a theory that is simple (though perhaps mathematically sophisticated) yet "explains" everything around us, every physical observation, then we are surely pleased with ourselves. (A Nobel Prize probably awaits.) However, can we be sure that our theory has anything to do with REALITY? Well, what is REALITY? Although our theories contain the beautiful logical connections we have already mentioned there is no demonstrable *logical* connection between any theory and the REAL WORLD. So *how* can we know what REALITY is? Yet, consider how we behave in everyday life: we do believe that we are dealing with *real* water when we drink it, wash with it, boil it, freeze it. We *operate* as if it is real (and we are real as well). We rely on its wetness (at usual temperatures) and all of its other properties, which we have experienced many times. We are assured by the fact that everyone else has had those same experiences with water. The key word in all of this is "operate." The experiments we set up to test our theories are analogous to (but more detailed than) such everyday experiences with the world, that have led us to form our ideas about the physical world. The scientific enterprise is therefore an extrapolation of the interweaving we have performed of our thoughts and our experiences to produce our common understanding of the stuff around us. This reality is referred to by philosophers as "operational reality." The *ultimate* question of what REALITY is does not concern us ordinary mortals, but is best left for philosophers, prophets, and poets!

LIGHTBULB INTERLUDE

So, how many physicists does it take to change a "lightbulb?" It all depends on those involved:

Scenario A

1. First a theorist creates the concept of a "lightbulb" as the source of the light and writes a paper which is published in a prestigious physics journal.
2. A theorist who enjoys building mathematical models reads the article of theorist 1 and is inspired to create a description of the structure of a "bulb." She/he writes a paper describing the description.
3. An experimentalist or theorist (phenomenologist) hears about "bulbs" and reads their proposed description. He/she then determines where and how to look for "bulbs." Another paper is written!
4. An experimentalist builds the equipment necessary to find "bulbs" and determine their detailed structure.
5. The experimentalist proceeds to search for "bulbs" and, after finding some, determines their properties. These findings either verify or modify the mathematical description. Several research papers result, some experimental and some theoretical.
6. An experimentalist unexpectedly finds a *dark* "bulb." A press conference and a celebrated research publication follow. (Perhaps a Nobel Prize is in the offing.)
7. An experimentalist or theorist then figures out, from the data found in steps 5 and 6, how to remove the dark "bulb" and replace it with one that will produce light.
8. Another experimentalist (or one of the above) finds a source of replacement "bulbs" that will glow.
9. An experimentalist *with unusual dexterity* removes the dark "bulb" and replaces it with one that does glow.

AND THEN THERE WAS LIGHT!

Scenarios B, C, ...: Permutations and/or elaborations of the above steps. Many of the steps may require the collaboration of several (to many) physicists, so that there is *uncertainty* and the number of physicists it takes depends on how we choose to *measure* that number.*

*You may want to look back at this last sentence after you have read about quantum theory.

Chapter 2

THE MATTER WE KNOW — PROPERTIES AND PRINCIPLES

1. THE SCIENCE OF ELEMENTARY PARTICLES

Elementary particle physics is the modern name of the field that attempts to answer the questions: what are the basic building blocks of matter and what holds them together? In fact, we would like to know and understand *all of the basic interactions and particles*, as well as the other observable basic features of nature. In ancient times, Greek philosophers believed that everything was composed of earth, air, fire and water, and correspondingly, in China the fundamental entities were metal, wood, water, fire, and earth (soil). Over the centuries, alchemists' (mainly unscientific) attempts to alter matter, for instance to change lead into gold, yielded enough understanding to lead to the science of chemistry. By the end of the nineteenth century, chemists had determined that all of matter was composed of about 90 basic elements. The smallest bit of an element possible is one particle of it, called an *atom* (from the Greek for "that which is uncuttable"). An atom is so small that one hundred million of them lying side by side would measure only about one inch.

From the earliest concepts to the resulting periodic table of elements, many small steps had been taken in our pursuit of the fundamental building blocks of nature. You may be wondering how we will know when we have reached the basic building blocks? That question, as we shall see, has no clear answer. A seemingly less grandiose question is: Could there be as many as 90 different *fundamental constituents* of matter (which the chemists have named "elements")? Well, why not? It really is a matter of taste — many scientists do not find it *aesthetically* pleasing to have so many; instead, they seek an underlying simplicity and "elegance." Well, nature may say "too bad!" BUT: Before 1900, it had already been shown that there existed a very light (as opposed to heavy) particle which is a common *constituent* of atoms, namely the *electron*. (These are the particles that flow in wires when we use electricity in everyday life.) The first elementary particle had been discovered... *and then there was one.*

Early in the twentieth century, a British physicist (from New Zealand) named Ernest Rutherford created and performed classic scattering experiments, in which he aimed electrically charged particles emanating from certain radioactive elements at various atoms. From the results of these experiments he deduced that most of the mass of any atom resides in a very tiny electrically charged central *nucleus* (roughly one one-hundred-thousandth the size of the atom). Imagine a hard candy, like those used to make lollypops. Now suppose that it has that fluffy cotton candy, sold at carnivals, all over it and extending out a distance one hundred thousand times the size of the hard center. If the hard candy were an inch across, then the cotton candy would extend more than three-quarters of a mile in all directions (or, in metric units, for 1 centimeter across we would have half a kilometer of cotton candy in all directions). This is very much like an atom, where the hard tiny center is the nucleus and the fluffy outer part is made of electrons in orbit. The diffuseness of the cotton candy is, in fact, a good representation of the quantum-wave nature of the electrons. Actual atoms are about a trillionth* the size of this candy model. Looking at this model from the outside, you cannot tell that it has a hard heavy center; furthermore, if this candy treat were extremely small, we could not poke our fingers into the cotton candy to explore its structure. In fact, we would not expect there to be a hard center inside. One way to find out about the structure would be to send tiny surrogate investigators in — namely, to fire tiny pellets into it; we would expect them to go through with some small deviations due to the structure of the cotton candy. What Rutherford found was that some of his (electrically charged) "pellets" did not go all the way through, but instead scattered backwards. That backward scattering was the first evidence of the existence of the extremely small, electrically charged nucleus. Further experiments found the actual sizes of the nuclei. The lightest element (hydrogen) has a singly charged nucleus. That nucleus is considered a single nuclear particle known as the *proton*. Perhaps this heavy minute particle was the *basic constituent* of the larger and heavier nuclei... *and then there were two.*

However, logical difficulties arose in attempts to understand the many different nuclei as combinations of only protons and possibly electrons. In the early 1930s, an electrically neutral (uncharged) particle was discovered which was both heavy and tiny, just like the proton. Could this be merely a composite, made of a proton and an electron, the same structure as the hydrogen atom but *one one-hundred-thousandth* the size? The answer was found to be a resounding NO! So this was a new elementary particle — its name? — the *neutron... and then there were three.*

Physicists often call a particle "elementary" when we know nothing about its structure, i.e., this appellation sometimes expresses our ignorance! Towards the end of the nineteenth century, atoms were thought of as elementary, and to this day are still called "elements" — the study of elements and their combinations (compounds),

*10^{-12} times the size of this model.

namely chemistry, can be thought of as the "elementary particle physics" of that era. The discoveries of the electron and the atom's nucleus and the study of their mutual interactions led to our understanding of the quantum behavior of nature, and that era's "elementary particle physics" has developed into atomic physics, quantum mechanics, and quantum chemistry. The study of the structure of nuclei, made of protons and neutrons, was the "elementary particle physics" of the 1930's and 1940's and has evolved into both nuclear physics and nuclear chemistry. On *our* journey, we will explore the secrets uncovered by the modern "elementary particle physics" that developed beyond nuclear physics.

I have not mentioned the many ways in which chemistry and nuclear physics have entered our lives as products, medicines and medical treatments, energy sources, etc., with their concomitant benefits and dangers. This is because we are dealing here with science and not technology. Technology, at its best, employs scientific knowledge and tools to enhance our lives. Sometimes it even creates better tools with which to coax more secrets from nature. However, it is not a part of science (from the Latin *scienta* — knowledge), although there is a natural symbiotic relationship between science and technology.

We have used and will continue using the word "particle" over and over again. Among the definitions given in *Webster's Encyclopedic Unabridged Dictionary of the English Language* we find: "one of the extremely small constituents of matter as an atom or nucleus... a body considered as possessing finite mass but infinitesimal dimensions... a body in which the internal motion is negligible...." The last of these is the sense in which we use the word "particle." (We shall see that the "infinitesimal dimension" property is mitigated by the quantum behavior of matter.) When we are studying a gas, the individual atoms of that gas are generally treated as immutable particles; when studying the atom, the electron and the nucleus are the particles being considered; when studying the nucleus, the constituent protons and neutrons are treated as elementary particles.

Many types of particles that no longer exist naturally on Earth have been created in high-energy collisions since the 1940s. Those particles are referred to as elementary particles because they, along with the proton and neutron, were originally thought (through the1960s, by most physicists) to have no constituents within them or, perhaps, to somehow be compounds of one another. In the 1970s, it was shown that most of these particles are, in fact, composed of more fundamental constituents called "quarks" (the vowel sound is the same as in "quart").

We shall soon discuss the fact that every particle has a corresponding antiparticle; some are in fact their own antiparticles. An antiparticle is almost identical to its corresponding particle but with opposite signs for electric charge and other charge-like properties. In each case, the antiparticle of an antiparticle is the original particle, just as I am the spouse of my spouse. Just as the word "man" sometimes has been used in the generic sense to represent all human beings, both men and women, the word "particle" or the word representing a particular type of particle is often used in the same generic sense to represent both particle and antiparticle.

Let us take a first look at what we have learned by observation about the particles of matter and their interactions. In later chapters we will tour the theoretical ideas, which "explain" these observations, and the experimental bases and verifications of those theories.

2. GENERAL PROPERTIES OF PARTICLES — A PRELIMINARY LOOK

Particles are identified and distinguished by their *masses, charges and spins* as well as by other more arcane properties that will be discussed in appropriate contexts. Some particles exist for only a finite time and then change ("decay") into other types of particles. For those decaying particles, their "lifetime" is a significant property as well. Let us first become acquainted with the simple general properties of mass, charge, and spin of particles as well as antiparticles.

Mass

You might expect that an important property of a particle would be its weight. However, if I would order my usual lox and bagel sandwich at the Lunar Cafeteria (on the moon) by asking for 1 ounce of lox, I would get 6 times as much lox as I like — weight is the pull due to gravity, and so it varies with location. If instead I asked for 28 grams of lox, I would be happily munching my sandwich at my favorite table near the window. That is because grams are a unit of mass, which is a definite *amount* of a substance. Consider first how we determine weight. We put our lox on a scale, which shows how hard the pull of gravity is on it. On the moon the pull on 28 grams is only 1/6 of an ounce. It takes six times as much lox to get a one-ounce pull. And the rule of thumb that one kilogram (or "kilo" for short) is the "same as 2.2 lbs." would not be true on the moon. A kilo *weighs* 2.2 lbs. on the Earth, but it would weigh only 1/6 of that, which is *a little less than 6 oz, on the moon*. To determine whether we have 28 grams, we can *compare* at the same location the pull of gravity on our lox to the pull of gravity on a standard set of masses that total 28 grams (perhaps a 25-gram, a 2-gram, and a 1-gram mass). We can use a seesaw type of arrangement, technically called an equal-arm balance, with the lox on one side being balanced by the masses on the other. This comparison to standard masses is analogous to the way we measure distances by comparing them to the standard distances marked on our rulers.

Charge

If you rub a balloon on your clothes it will stick to the ceiling. Comb your hair when it's dry and the comb can pick up small pieces of paper. If your shirt or blouse contains polyester or Dacron you may notice that it crackles as you crumple

it after taking it off; in a dark room you may notice flashes of light as you crumple it. These are examples of *static electricity*. It has been found experimentally that rubbing two different materials together leaves them in a markedly different state. We can charge a small object with static electricity by rubbing a rubber rod with fur and then touching the object with the rod. We observe that two objects charged in this way will move away from one other. — They exert a repulsive force on one another. A similar experiment can be done with a glass rod and silk. However, if we charge one object with a glass rod and another with a rubber rod they do not repel but move towards one another. They are attracted to each other. This tells us that they must contain *different kinds* of charge. When we put the same amount of these two different kinds of charge on an object, they appear to cancel so that the object behaves as if it were uncharged. From such experiments and other more detailed investigations we learn that there are two kinds of electrical charges which can be named positive and negative charges. Charges of like sign repel whereas those of opposite sign attract. (Remember that old adage "Opposites attract!")

The word "static" in "static electricity" means charges that are not moving. Moving charges have proven to be of greater interest and utility. A dramatic example of natural phenomena that involve moving charges is lightning, which is a flow of electrical charges between the sky and the earth. Over the past couple of hundred years, we have learned how to cause and employ flows of electric charge (in wires, etc.), which we refer to as electricity or electric current.

Actually, the choice made by scientists long ago about the direction of flow of electricity leads to the conventional negative sign of the electron charge. At that time no one knew about electrons, which are the charged particles moving when current flows through wires. The sign convention chosen leads inadvertently to the fact that when we say that electric current is flowing in a wire from A to B, electrons are actually moving from B to A.

<div align="center">

Flow On

(with apologies to the composer of "Glow Worm")

Flow you electrons,
down the wire.
Do not be worried;
do not tire.
Although we *say* that *currents* flow,
your *opposite* course we know!

</div>

If we had a magnet sitting near a static charge, we would not see any interaction between them. The magnet will exert forces on other magnets, on iron, etc., but not on a static charge. And so, magnetism was thought to be independent of electrical phenomena. However, moving magnets do exert forces on static charges. Similarly, flowing charges exert forces on magnets and also experience forces in the presence of magnets. These *electromagnetic* interactions show that the electrical behavior and the magnetic behavior of matter are intertwined. In fact, the

ubiquitous electrification accomplished during the first half of the twentieth century was based on electric generators and electric motors, which exploit this interplay between electricity and magnetism.

These applications of electromagnetism have radically transformed our everyday lives — our achievements, our ambitions, and our dreams. The fundamental concepts that made all of this possible were developed by scientists over a period of two centuries. The results of their obsessive curiosity and meticulous patience — also known as *basic research* — have wrought the magnificent technological revolutions of the twentieth century, as well as enlarged and deepened our understanding of the universe.

<u>Electricity</u>

EE	ee
ll	ll
ee	ee
cc	cc
tt	tt
rr	rr
ii	ii
cc	cc
ii	ii
tt	tt
yy	yy

**Although we're surrounded by electricity,
we're no longer astounded by electricity!
We use it with nary a thought,
and no explanations are sought.
'T has transformed our days
in such marvelous ways;
technology vibrant and strong
that arose from our science's song.
By more than two centuries' thought,
this wondrous advancement was wrought.**

———— **P l u g it in !!** ————

(If you have not already done so, please look at this page from a distance.)

Spin

An object that is not moving from one location to another may still be spinning about an axis just as the Earth does. This behavior is described as *angular* motion. The rapidity of the angular motion is characterized by a quantity called "angular momentum." We shall soon see that particles in motion have another type of angular momentum, so to distinguish we refer to the angular momentum of a spinning object as "spin." An elementary particle behaves roughly like a spinning object; however, we shall see later that that picture is not strictly true, due to the quantum nature of matter. Nevertheless, even though an elementary particle is *not* a chunk of spinning material, it does behave as if it *were* in the sense that it has *intrinsic* spin.

<div align="center">

A Point Particle?

Its infinitesimal size:
impossible to visualize.
But quantum mechanics can show:
that particle 's never quite so.
And how can a point have a spin?
Intrinsic 's the answer we're gi'n.

</div>

Quantum theory also shows us that the spin of a particle cannot be just any amount. Instead, each particle's spin must be an integral number of spin units, just as the prices of items we buy are an integral number of cents — no fractions of cents are allowed. When the allowed values of a quantity are discrete rather than continuous, we say that the quantity is quantized. Therefore our money and the spin of particles are both *quantized*. When you enter a building that has steps and also a ramp leading to the front door, if you take the ramp you are going up a structure with *continuous* levels, if you walk up the steps you've chosen the *quantized* approach.

There is a basic unit for angular momentum just as a penny is the basic unit of US money, but its value is not important for our discussion. In terms of that unit, the lowest possible spin is 0, the next lowest is 1/2, then 1, etc. All particles fall into one of two categories based on spin. A particle whose spin is an integer is called a *boson*; a particle whose spin is an odd half-integer is called a *fermion*. Electrons, protons and neutrons have spin 1/2, so that they are fermions.

Now let us take our first look at the electromagnetic interactions and the associated electromagnetic waves. Electromagnetic waves include radio waves, microwaves, infrared waves, visible light, ultraviolet waves, x-rays and gamma rays. We shall see later that these waves are actually made up of quantum particles we call "photons," which are spin 1 bosons. (In fact, we have just listed these waves in increasing order of the energy of the corresponding photon.) Furthermore, we will see that the electric and magnetic interactions between particles are also understood to involve these spin 1 photons.

In fact,

- *all of the particles that we believe* to be the fundamental building blocks of matter have spin 1/2 so that they are fermions,*
- *and all of the particles associated with the interactions between matter particles have spin 1 or 2 so that they are bosons.*

<div align="center">

Quantized Spin

Yes, particles have spin they say,
a property that's hidden.
The mathematics shows the way
most values are forbidden.

The steps of quantized values found
have size that's just one-half;
But nothing actually spins around;
there is no photograph.

And yet we've found the spin of every
particle we know.
With pride, our ingenuity
has caused their spins to show.

The fundamental *matter* ones
have spin one-half, we've shown.
So they are surely fermions,
from theory it is known.

And interactions' delegates,
that play their roles on cue,
the photons and their boson mates
have spins of one or two.

Spin, spin! Glorious spin.
You look at us with that sly grin.
Your mystery won't do us in!
Our math reveals your yang and yin!

</div>

A spinning ball of charge has its distributed charge in *motion*, so that it has magnetic properties just like those described above for *moving* charges. And a charged particle with *intrinsic* spin has electromagnetic properties similar to those of a spinning ball of charge. As a result of this interplay between electric and magnetic phenomena, it behaves very much like a magnet. In fact, the underlying magnetic properties of magnets are understood to be due to the motions around the nucleus and the spins of the electrons.

Antiparticles

For *each* type of particle there exists an antiparticle (the first *i* is pronounced 'ee') of the same mass, lifetime (if it is unstable), and spin, but with a charge of opposite sign

*Whether they will be found to be composites of yet unknown entities is an open question.

(likewise for other charge-like features to be discussed later). The first antiparticle found was the *positron*, which is the antiparticle of the electron (e^-) and is usually designated e^+. The antiparticle of the proton is called the *antiproton* (designated \bar{p} and called p-bar), that of the neutron is the *antineutron* (\bar{n}) which we shall see is different from the neutron even though they are both electrically neutral. The antiparticle of the photon is the photon itself, a fact that is sometimes expressed by saying that the photon has no (distinct) antiparticle. In each case, the antiparticles of the antiparticles are the corresponding particles, so that the antiparticle of the \bar{p} is the p, that of the e^+ is the e^-, etc.

The part of our universe that we know, including the solar system, the Earth, this book, and you, dear reader, are made of protons, neutrons, and electrons. However, we could imagine a universe that is constructed in exactly the same way but with antiprotons, antineutrons, and positrons; we use the word "antimatter" to describe matter made of the antiparticles of our *ordinary* matter. Researchers have succeeded, so far, in constructing anti-hydrogen, composed of an antiproton as its nucleus with a positron going around it.

When a particle interacts with its antiparticle, the result is usually mutual *annihilation*. Both particles are destroyed with the concurrent *creation* of other particles. For example, when a positron and electron annihilate, photons are usually created in the process. Even in its earliest form, the theory of quantum electrodynamics, which contained electrons and photons, predicted the existence of antiparticles (in particular of the positron) *before* their discovery. It also correctly predicted the *annihilation* of a positron and an electron accompanied by the *creation* of photons.

<u>And Then There Were Photons</u>[*]
An electron, while trav'ling in space,
met a positron there "face-to-face."
The electron then sighed,
at the sight of his bride,
and they "died" in a loving embrace.[†]

In summary

These general properties — *mass, charge, spin and lifetime (for unstable particles)* — are parts of the IDs of the elementary particles of nature. Now we turn to their motional (without an "e" at the beginning) behavior.

[*]This limerick won a prize in the contest held by the American Physical Society, and appeared in the March '97 issue of the APS NEWS.

[†]To my Asian friends and others whose languages have "negative" as feminine, I wish to point out that the electron is considered part of "matter," and the positron (its antiparticle) part of "antimatter." When we learn about Lepton Number, we will see that the electron has +1 for its value and the positron has −1. We could even say that Ben Franklin and the others, who *chose* which kind of charge would be called plus and which minus, made the wrong choice. (We have already seen that the electrons are usually the mobile charges when currents flow).

3. PARTICLES IN MOTION

Momentum

We use the word *momentum* to refer to the motion of the particle from one place to another; technically this is called *linear* momentum as opposed to the *angular* momentum that we have already described. A particle sitting still has no momentum, whereas a particle in motion has momentum, which is greater the faster it is going. When a bowling ball bangs into your foot and stops, you feel quite a jolt! (Ow!) But if it had been a ping-pong ball instead, you may not have noticed at all. This suggests that the bowling ball had a lot more momentum than did the ping-pong ball. It is a dramatic demonstration of the fact that momentum depends upon mass as well as speed; the more massive object had more momentum although they were moving at exactly the same speed.*

Actually, a full characterization of momentum requires the designation of the *direction* of motion as well. The momentum of an *isolated* particle, neither momentum size nor direction changes as it moves; it takes a push or pull (*i.e.*, a force) to change its momentum. This is a modern version of the law of inertia.

In a reaction among particles (with no forces from sources outside those particles), the sum of the momenta of the particles remains unchanged. This is the principle of *conservation of momentum*. As one particle feels a push from another particle, the second one feels the same size push from the first such that their respective changes in momentum are equal but in *opposite* directions; they therefore cancel in the mathematical sum. If you ice skate, then you are probably aware of the conservation of momentum. If not, start by standing still on the ice behind a friend; now, as you push your friend forward, you will be thrust backward — conservation of momentum! This doesn't work on the floor, because friction holds you fixed to the floor; it is the floor or the building, or perhaps a segment of the earth's crust that recoils by an extremely tiny (unmeasurable) amount.

More generally, if we take the entire universe as our "collection of particles," we conclude that the total momentum of the universe is conserved.

Orbital Angular Momentum

The motion of an object following a closed path, returning to its original position again and again, may also be described with the language of angular motion just as spin is; the Earth's motion around the sun falls into this category. This *orbital* angular momentum is quantized just as is spin, but the allowed values are only integers $(0, 1, 2, \ldots)$ times the basic unit of angular momentum.

*The size of the momentum is the mass *times* the speed.

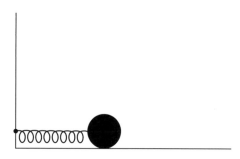

Fig. 2.1. Particle attached to stretched spring. The spring is anchored to the left wall. The floor
is so smooth that we can ignore the effect of friction as we watch the particle slide.

Energy

A particle sitting still has no motional energy, which we call "kinetic energy." The
kinetic energy of a particle depends on its speed and mass or may be determined
from its momentum and mass. When the momentum is zero it is sitting still, so
that its kinetic energy is zero as well.

Suppose that a particle were attached to the end of a stretched spring (anchored
to a wall), as shown in Fig. 2.1. If we let the particle go it would speed up as the
spring pulled it toward the wall. We can describe this situation by ascribing energy
to the particle due to its *location* at the stretched-spring position.* This energy of
position is referred to as "potential energy." We can think of the potential energy of
the particle as energy "stored" in the spring. As the particle speeds up, its kinetic
energy increasing, the spring becomes less and less stretched so that the particle's
potential energy is continually decreasing. However, the sum of the particle's kinetic
and potential energy, i.e., its total energy, remains the same throughout the motion
of the particle.

There are other forms of energy besides the two described here. The only other
form that must be considered when discussing small numbers of particles is called
the *rest energy*. An isolated particle at rest possesses no kinetic energy or potential
energy; nevertheless, it has energy due to its mass, which we call "rest energy." It's
the old "$E = mc^2$." This is one of the important things we learn from Einstein's
theory of relativity.

When we take into account all the different forms of energy, we find that, for
any interaction, the total energy remains unchanged. This is the principle of *con-
servation of energy*. In the example of a particle at the end of a spring, the kinetic
and potential energies were changing, with their sum remaining the same. In the
fission of uranium that powers nuclear reactors, it is actually mass (or rest energy)
that drops, with a corresponding increase in kinetic energy. This leads to an the
increase in the random motion of the molecules of water used for energy transfer

*An alternate description of this process involves the concept of work, which we choose to omit.

in the reactor, as they are hit by the fast-moving fission products; that increase in random motion corresponds to an increase in the water temperature. This kind of transferred energy is called *heat*. This kind of transferred energy is called *heat*. (In ordinary (macroscopic) systems with moving parts, friction appears to lower the energy. Actually, the friction only *changes* some of the motional energy of a system into heat. The *total* energy remains conserved.)

We now explore the concepts of relativity that are pertinent to our understanding of particle interactions. These considerations are known as the Special Theory of Relativity. Our modern understanding of gravity, cosmological structures, and the evolution of the universe come from General Relativity, which contains a much deeper understanding of the nature of spacetime. However, the gravitational interaction is very weak, and does not play a significant role in interactions among small numbers of particles. This allows the independent exploration of the other interactions of nature without the complications that General Relativity would require. Consequently, we will not discuss the beautiful but arcane theory of General Relativity.

Chapter 3

THE RELATIVITY OF SPACE AND TIME

1. THE PRINCIPLE

Imagine a child living near a straight, flat highway. His father has a car whose engine is inaudible. While sitting in the back seat, the child is throwing a ball up and catching it, waiting for his father to drive him to the game. It's a long wait and the child eventually falls asleep. After awaking the child resumes throwing the ball up and catching it. He is so intent on this activity that he fails to notice whether his father is in the car or whether the car is moving. Could this child tell from the behavior of the ball whether the car is moving along the highway at constant speed or is still parked in front of his house? The answer is no! Have you ever dropped something in a car or a plane while it was traveling at a constant speed and direction? If so, didn't it fall straight down? Dropping an object or throwing it up and watching its motion is, in fact, a qualitative *experiment* that studies the motion of objects and the effects of gravity on them. These experiments and their quantitative elaborations, which are our *usual* experiments, show that the motion of objects as well as gravity itself obey the same laws in a vehicle moving at constant velocity as they do when the vehicle is sitting still.

The type of vehicle (car or plane) of course has nothing to do with the motion of the object, and we replace any reference to it by the words "reference frame." By *reference frame* we mean any framework at rest with respect to the observer that may be used to determine the location of objects as well as the times at which the observations are being made. A reference frame is actually not massive like the interior of the vehicle, but consists of hypothetical space-axes and clocks. An observer (in this case the boy) is, by definition, at rest in his own reference frame. It is natural for us to think of ourselves as not moving, or at rest, when we are sitting in our easy chair. This is because we are at rest with respect to the largest object in our vicinity, namely Earth. Similarly the kid in the car, which was in fact moving rapidly along the highway, was at rest with respect to the car — so that *his* frame of reference coincides with the car. Consider two starships that are not near

any large body. If the starship E and the starship H are moving past one another (engines off), the crew members on each believe that *they* are at rest and that those on the other ship are moving. They are *all* correct! Those on the starship E are at rest with respect to their surroundings — essentially the starship E, and similarly those on the starship H are at rest with respect to their surroundings — essentially the starship H.

Now suppose that during the drive the kid's father hits the accelerator when the ball is at the top of its path. This causes the car's reference frame to *accelerate* with respect to the reference frame of the highway. The ball in the air will *not* accelerate, because it is not attached to the car in any way. However, you know that the little guy will feel that acceleration in his whole body as he notices the ball shooting backwards. Actually, he will feel the seat pulling him forward (keeping him in the car's accelerated frame); however, his internal organs are not quite keeping up. How do things look in this accelerating frame of reference? It is apparent that the law of inertia and the conservation of momentum are not true in this accelerating frame because loose objects (on the seats or the floor) start to slide backwards when the accelerator is pushed down. The ball in the air, which originally had no horizontal momentum in the kid's frame of reference, acquires momentum backwards in this accelerated frame. The momenta of all these objects change as if there were a force, called an *apparent* force, on them. (These apparent forces are sometimes called "fictitious" forces instead because they are simply the result of being in an accelerated frame.) The overall conservation of momentum, as well as the law of inertia and Newton's law connecting momentum change to applied forces, are modified in an accelerating frame. The effects of that acceleration make things behave, in that reference frame, as if there were forces (appropriate to the frame's acceleration) on everything. When test tubes spin about in a centrifuge, heavier particles move outward; when we go on the rotating carnival ride (where the floor drops away), we feel an outward force that presses us against the outer wall, whose friction keeps us from dropping when the floor drops. These are examples of centrifugal force, which is a commonly encountered apparent force. Centrifugal forces are felt in rotating frames of reference because they are *accelerated* frames.

Those frames of reference in which the law of inertia and the conservation of momentum are true are referred to as *inertial frames*; in these frames there are no *apparent* (fictitious) forces. What we have seen here is that any frame (like the moving car before the accelerator was hit) moving at a constant velocity with respect to an *inertial* frame (like the car at rest) is also an inertial frame. This is one version of the *principle of relativity*, which says that the laws of nature are the same in all frames moving at constant velocity with respect to one another. Therefore, there is no frame of reference that can be designated "truly at rest," nor can we characterize frames as truly moving. All observers (whose rest frames are inertial), and their respective reference frames, are equally *worthy*.

Note that we have assumed here that the law of inertia holds in the reference frame of the earth and anything sitting still on it. However, this is not precisely true due to (i) the earth's spin about its axis and (ii) the earth's nearly circular motion around the sun, which itself is moving through the galaxy. But the law of inertia is very nearly true on the earth, so that using it as our representative inertial frame is appropriate for our discussion.

Now let's look in on the little guy throwing the ball up and catching it in the "moving" car. His cousin is sitting on the side of the road watching. Before the accelerator is hit, she sees everything in the car moving at the same speed as the car itself. The thrown ball is moving down the road at the same speed as the car while it is going up and down, so that it *does* keep up with our hero's hands and can be subsequently caught. When the car accelerates, however, the ball is in the air (not in contact with any part of the car) so that it keeps moving horizontally at the *original* speed of the car. But the car and the boy in it are moving faster, so that he sees the ball moving backwards. His cousin is in an (essentially) inertial frame, and since the ball in the air has no *actual* horizontal force on it, its horizontal momentum is unchanged in her frame of reference. The apparent (fictitious) force on the ball perceived by the boy in the car is *not* the result of any interaction; it arises solely from the acceleration of his frame of reference (which is *not* an inertial frame during the acceleration) — the cousin sees the *boy* accelerating past the *ball* in the air, whereas the boy sees the ball accelerating back towards the rear window.

From now on, we shall talk only of inertial frames to avoid the extra complications that arise in frames accelerating with respect to them. When we are at rest in an inertial frame, our stomachs do not feel like they do when we are coming down the big hill on a roller coaster or even going up — there are no *apparent* forces. In these frames, all changes of momentum result from actual physical forces exerted by other objects.

Let us consider the "experiment" being done by the boy, in his *inertial* frame, when the car was moving at *constant speed*. He threw the ball up and it came straight down. It slowed as it went up, stopped, and then continually increased speed as it came down. This was due to the pull of gravity. The cousin on the side of the road watching that experiment saw a more complicated behavior with horizontal motion accompanying the rise and fall of the ball. She could, however, do an experiment identical in her inertial frame to what the boy was doing in the car by throwing a ball straight up at exactly the same speed as he did. If they each made careful measurements of their own ball's motion, the results would be identical within experimental precision, assuming no wind or other disturbances. Thus in identically constructed experiments, they would obtain identical results finding the *same* laws of nature in their *different* inertial frames — another way of expressing the principle of relativity.

2. ELECTROMAGNETIC WAVES

It was discovered during the 19$^{\text{th}}$ century, by James Clark Maxwell, that the laws of electromagnetism account for the existence of electromagnetic waves and reveal that ordinary light is such a wave. These laws show that those waves move at a speed of 186,000 miles per second (300,000 kilometers per second), that is about 670 million *miles per hour* (one billion *km per hour*). This (enormous) speed of light is usually designated by the letter c.

Fig. 3.1. The starship E emits an electromagnetic wave towards the starship H, which it sees moving away at a speed of 2/3 c. The question is what is the speed of the electromagnetic wave read by the sensors in *starship H?*

Now, suppose that a radio message, which is an electromagnetic wave, is sent out by our starship E to the starship H. Our sensors would show the wave moving away from us at speed c. If the starship H is moving away from us at a speed of two thirds of the speed of light (2/3c) (See Fig. 3.1.), wouldn't its crew see the wave approaching at one third of the speed of light (1/3c)? Certainly, if I am (lawfully) traveling at 65 mph and the car ahead is traveling at 64 mph, in that car's frame of reference I will be approaching at 1 mph. So, suppose that the answer were yes; the radio wave was approaching the starship H at 1/3c. What's wrong with that? It seems simple enough — no big surprise. However, what about the principle of relativity? If the same laws of electromagnetism hold in the frame of reference of the starship H as hold in that of the starship E, the electromagnetic wave should be traveling at speed c according to the sensors of the starship H. But if it were moving at speed c to those in the starship H, wouldn't it have to be going at $1\frac{2}{3}c$ in the frame of reference of our starship E? So which starship "wins the award," which has the electromagnetic wave moving at c?

When the laws of electrodynamics were completely understood, it seemed apparent that they were not consistent with the principle of relativity: The speed of an electromagnetic wave was predicted to be c by the equations, but even in the simple scenario above it *shouldn't* be c in both frames — if it is c in one, then it should not be c in the other. So it was believed that there was a frame which was *truly* at rest, in which the speed of electromagnetic waves *was* c, and this was referred to as the "ether frame." The name arose from the belief that an electromagnetic wave was an oscillation of hypothesized matter called "ether" that pervaded all of space (even solid glass). However, experiments to detect the ether found *nothing*, showing that the electric and magnetic disturbances *themselves* are the oscillating components of the wave.

Just after the turn of the twentieth century (1905), Einstein expressed his firm confidence in the principle of relativity by demanding that the principle apply to electromagnetism as well. His demand would imply that, although the radio wave was traveling away at speed c toward the starship H (moving away from the starship E at speed $2/3\ c$) the speed of the wave according to the starship H was also c — not $1/3\ c$ as we would expect. Well, how can that be? It defies simple intuition!

<div align="center">

Race Not, Lose Not

To race 'gainst a photon 's a futile activity
No matter how fast you may be.
We learn from the theory that 's called relativity
that it will go past you at c.

</div>

3. SPACE AND TIME

In our discussion of the two starships in relative motion, it appeared that there was no way that a wave (or anything else) sent from one to the other, could be going at the same speed according to both ships — unless our understanding of space and time was *wrong*, since speed is simply distance covered divided by the time interval of the motion (as in miles per hour). In order to explore the kinds of changes required of our fundamental understanding of space and time, Einstein created some incisive *thought experiments*, which could be done in principle but were not necessarily practical. These simple conceptual experiments revealed unexpected properties of space and time themselves as well as relationships between them. Let us examine a few simple thought experiments based on those created by Einstein.

Time dilation

Suppose that, on top of our starship, we constructed a simple clock using light. In this clock a pulse of light travels to a mirror and reflects back to a detector on the laser that emitted it. This arrival causes the clock to "tick" and also triggers the emission of the next pulse. For a length of 750 meters, two hundred thousand round-trips take one second (click). Voila — a clock! Suppose a starship is moving by, in a direction perpendicular to the path of the light pulse, and watching our clock. Their crew members see us and our clock in motion, in the opposite direction to their motion with respect to us. They detect *the same two events*: (1) the pulse leaves the laser, (2) the pulse arrives back at the laser. But in their frame the laser has moved sideways (along with our ship) in a direction perpendicular to the line from the laser to the mirror, as shown in Fig. 3.2.

The light also is seen by them to have this same sidewise motion, so that it keeps up with the sidewise motion of the laser and finally reaches it. So, in their frame of reference, the light traveled a *greater distance* than it did in ours; it covered the distance between the laser and the mirror, up and back, as well as the sidewise

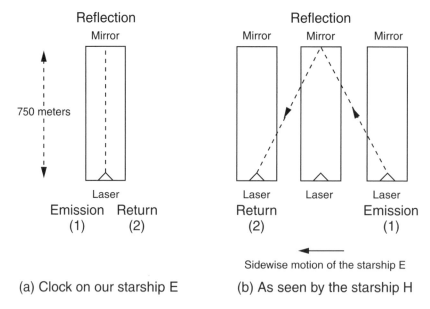

Fig. 3.2. (a) Our clock, as seen by us, with the light path shown as a dashed line. Emission and return are at the same place. (b) Our clock, as seen by the crewmembers of the starship H, which is moving by us to the right. To them, the clock is in motion (along with our starship) to the left. Here we show its position at three different times: at the time of emission (1) of the light, the time of reflection, and the time of return (2). The (longer) light path is again shown as a dashed line.

distance moved by the laser.* If light travels at the *same* speed in all frames of reference, it should take a *longer time* to return to the laser according to the clocks in starship H. Its crew would say that our clock is *slow*. On the other hand, if they have a clock just like ours mounted on *their* starship and we watch the two events: (1) the pulse leaves *their* laser, (2) the pulse arrives back at *their* laser, we would say that *their* clock runs slow (by the same reasoning). Which is it? Even though it may appear paradoxical, both are correct! This is one example of the fact that time intervals, the time between two events, are not absolute in a *relativistic world*.

The two events at our laser occur at the *same place* in our frame of reference. In a reference frame that sees the two events at the same location, we call the observed time a *proper time*. The other ship sees those two events occurring at *different places* so that the time interval between those events is not a proper time in their frame of reference. We have just demonstrated that *the proper time is the shortest time*, since *the distance traveled* by the light is *shortest* in the frame of reference for which the events are at the same place. Thus, observers in other frames measure a longer or *dilated* time. This is also true when we look at *their* clock — to them

*It requires only elementary geometry (the Pythagorean Theorem) to figure out from Fig. 3.2 the ratio of the distance the light traveled in their frame to that in ours.

the interval between emission and return is a proper time, but not to us, so we say that *that* interval was longer than one two-hundred-thousandth of a second. There is no paradox, since in the two clocks we are looking at different sets of events. In fact, there is a symmetry between the two starships, each saying exactly the same thing about the other one's "identical" clock. Here we see the equivalence of the two frames, at least in this respect.

<div align="center">

Relative Time

(with apologies to the composer of "Pretend")

Pretend you're early when you're late.
It isn't very hard to state,
you set your wristwatch in a reference frame
that's not the same as theirs.
And just in case they doubt your word,
you'll prove your claim is not absurd,
with special rela-tiv-ity
and *c*.

</div>

Suppose that *we* observe two events while the sensors on the starship H are not functioning. If we relay the readings on the starship E's sensors to the starship H, could its crew figure out what their sensors would have read? The answer is yes; however, if we told them only the amount of time we measured between the two events, without our measurement of the distance between those events, they could not figure out what time interval would have been read on *their* sensors. If the two events were at the same place in our frame, so that it was a proper time in our frame, their measurement would have shown a longer time. If it were not a proper time in our frame, it might not be as much longer and might even be shorter (certainly, if it were a proper time in *their* frame). *Our* measurement of the *distance* between the events is needed to be able to do the elementary geometry required to figure out the *time* interval in *their* reference frame. Other thought experiments show that similarly the space separation (distance) between the events cannot be figured out on the starship H unless we send them *both* the space separation and the time interval measured by our ship's sensors. This intertwining of space and time leads us to proclaim that we live in a *four-dimensional spacetime continuum*, rather than in a three-dimensional space.

The relativity of simultaneity

We have just seen that time intervals are not absolute. Now we will see that even simultaneity is not an absolute concept. Suppose *you are in a train car moving* at constant velocity, as shown in Fig. 3.3. There is a light source at the center of the car, which fires two flashes of light at the instant it is triggered, one forwards and one backwards. Since the front and back walls are the same distance from the

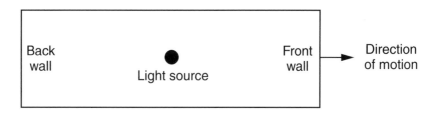

Fig. 3.3. Moving train.

source, the flashes reach those walls simultaneously. This is the way *you* experience it in your frame of reference (the train car). However, to someone on the platform, the *back wall* is moving *toward* the left flash and the *front wall* is moving *away* from the right flash because the entire train is moving. But since both flashes of light are moving at the same speed (c), the left flash hits the *approaching back wall sooner* than the right flash catches up to the *receding front wall*. Thus, although the two events — (1) left flash hits back wall; (2) right flash hits front wall — are simultaneous in the frame of the moving train, they are not simultaneous in the frame of the platform. So we see that *even simultaneity* is not absolute, but instead depends on the reference frame.

Length contraction

Suppose a person on the platform wants to measure the length of the train car while it is moving. One way this could be accomplished would be to have each end (front wall and back wall) of the car spray a bit of paint *at the same time* onto the track, and to then measure the distance between the paint marks after the train car has gone by. It is important that the paint be sprayed *simultaneously* since the train is moving. "Ay! There's the rub." What is simultaneous to you on the train is not to someone on the platform. Suppose that, to make sure that the sprays turn on simultaneously, *you* (on the train) place an electric eye on each wall, which will fire the paint spray when it receives a light pulse. You then trigger the light source, knowing that the flashes will reach the two electric eyes simultaneously. You would consider the distance between the marks on the track to be the length of the car. However, based on the foregoing discussion, we know that the person on the platform would say that the back wall sprayed paint *before* the front wall did; the front wall moved forward during that time delay, so that the marks are too far apart. He would claim that the length of the car is actually *smaller* than their distance apart — a *contracted* value compared to *your* measurement. Likewise, the measured length of any object *moving* in a given frame of reference will be *contracted* along the direction of motion, compared to its size in its own rest frame.

It's a weird world, after all

So, the properties of space and time are peculiarly *unintuitive* from several points of view. First, time intervals and distances are not absolute; their sizes depend on the reference frame in which the measurements are made. Second, space and time intervals are intertwined so that we actually live in a spacetime continuum. It should be noted that these features of spacetime are only noticeable when the relative speed between reference frames is about one-tenth the speed of light or greater.

<div align="center">Spacetime is Fine</div>

"I will make no accusations,
but your 'meter' stick 's too short!
So now all your observations
must be fixed before report.

Yes, I've read the true locations,
at the same time, of both ends.
Yet subtracting those notations
gives a *different* length, my friends.

But wait! Here is something *strange* now
as you throttle down your speed,
your rod's *length begins to change now*.
So *my* measurements do read.

There is something wrong with space then.
Oh, alas and oh, alack!
Or is faulty time the case, then?
Maybe *both* are out of whack.

So, let's break our rulers, *burn them!*
And of length we'll speak no more.
And our clocks, we must now spurn them;
fickle time 's a common *whore!*

Yes, let's *break* our rulers, *burn them!*
They're no use to us at all.
And our clocks, we'll *surely* spurn them;
spacetime makes no sense at all!"

"Do not worry 'bout spacetime now
(or whenever 'tis to you).
We are both correct, and here's how:
relativity 's the clue.

The electromag. equations,
in all *frames*, are just the same,
and contain firm implications.
Light's speed, c, they do proclaim.

So, don't break your rulers, use them.
For length's mystery 's unveiled.
And your clocks, please don't abuse them.
Spacetime's subtleties we hail!

No, don't worry more about it;
in frustration, do not scream!
Though your intuition doubts it,
it is part of nature's scheme."

Causality's immutability

What happens to causality in this world of relative time intervals? If you throw an apple at the teacher that hits the blackboard instead (whew), is there a frame of reference in which the apple hit the blackboard *before* you threw it? In other words, can someone moving by our classroom (i.e., in another reference frame) actually find that the splat was *before* the throw? That would surely violate our understanding of cause (throw) and effect (splat)! It is my great pleasure to inform you that you need not worry; the mathematics shows that there is no reference frame in which the time-order of cause and effect is reversed. For any two events that can be connected by an apple or anything else moving at a speed not exceeding the speed of light, the time-order is the same in all frames of reference — the cause always precedes the effect! Yay; *there is still sanity in the world!*

<u>The Case of The Relative Murder</u>

SUMMARY	Professor *Fail* was shot and fell in front of Physics Hall.
OF PROCEEDINGS	A car was going by just then, a gun protruding some.
TO DATE	The fatal wound destroyed his heart, as he began to crawl.
	There were three students *there* with him; with grief they're overcome.
	The driver of th' car 's the one who fired that dire shot.
	He's been arrested and is standing trial for this foul crime.
	Defense attorneys wish to show new evidence they've got;
	the jury has been asked to leave the room now, for a time.
DEFENSE:	My client pulled the trigger, judge, on that we all agree.
	But we intend to show right now, although the man expired,
	that video tapes made from the car can prove with certainty
	the body fell 'n that reference frame *before* the gun was fired.
PROSECUTION:	Objection, please your honor, there are witnesses who'll tell
	the gun was shot and *then* the victim, on his face, fell flat.
	Ballistics showed that gun did surely fire the fateful shell
	and there 's no reference frame in which time-order flips like *that*.
JUDGE:	Objection sustained; the tapes are inadmissible!

The judge knows her physics!

4. ENERGY AND MOMENTUM

Before Einstein proposed his relativistic ideas, our concepts of space, time and motion were consistent with our instinctive intuition. Energy and momentum had a distinguished place because they were conserved quantities. Both the energy and momentum of an object depend upon its mass and speed; but speed depends on position and time measurements that are quite different in the theory of relativity. The mathematics of relativity shows that the *effective* (relativistic) mass of a moving object increases with speed, getting larger and larger without limit (*i.e.*, approaching infinity) as we get closer and closer to the speed of light. Consequently, the momentum and energy also increase this way. This means that it takes more and more force to increase the speed of a massive object as it gets closer to the speed of light, and that the speed of light is unattainable. The value of the relativistic mass when the object is at rest is its ordinary mass, which is sometimes referred to as its *rest mass*. Actually, the relativistic mass of a free particle is in essence its energy. This is the content of the famous formula $E = mc^2$, where m, the relativistic mass, is multiplied by the square of the speed of light to calculate the energy of the particle. It is redundant to speak of relativistic mass *and* of energy; from here on, we will use the word "mass" to mean "rest mass" and will never talk of the "relativistic mass" again. (This is in accord with the usual jargon of particle physics.) We will express the increase of relativistic mass with speed by saying that the closer the speed of a massive particle is to the speed of light, the higher is its *energy*.

And what about the photons that comprise electromagnetic waves? They are moving at *exactly* the speed of light c! These are particles whose (rest) masses are *identically* zero. And the equations show that massless particles *must* be moving at the speed of light in all frames of reference — they have no rest frames! Furthermore, *their* momenta are not infinite nor are their energies. There are, in fact, several other types of massless particles in nature, which we will discuss later. All of these particles are always moving at that "magic" speed c, which is the same in all frames of reference.

Relative Intuition

Can you relate to relativity?
Or is stubborn rejection your present proclivity?
Does your int(u)ition have the reliability,
To buttress your deny-ability?

Interlude

THE "ILLIONS" AND THE "ILLIONTHS"

In the course of our journey, we shall encounter both very large and very small numbers. Most of us do not have a *gut* feeling about how large a million, a billion, or a trillion really is, nor the smallness of a millionth, a billionth, or a trillionth. Let us try to put these numbers into rough perspective.

A million is a big number. A thousand of those is called a billion, and a thousand billions is called a trillion (which is also a million millions). To get a feel for these enormous numbers, let us recall the children's game hide-'n-go-seek. In this game, the person who is "it" raises his/her arm (as if reading his/her wristwatch) to eye-level. He/She then leans that arm across a favorite lamp pole or tree trunk (with closed eyes against the raised arm) and counts to one hundred, while the others hide. Counting very fast, the count to 100 takes about 20 seconds. Suppose the count were to one thousand; it would take about 3 minutes and 20 seconds. To count to one million takes about 2 days and 7 hours — a long, lonely, hungry time. To count to one billion takes about 6 years and 4 months — by which time the kids would outgrow the game. To count to one trillion takes about 6,300 years — forget it!

The "illionths" are 1 divided by the corresponding "illions," so they are smaller than we can intuitively imagine. In order to appreciate their smallness, let us imagine that we have constructed a very accurate clock. Suppose it loses only one millionth of a second every second. Our clock would lose only one second about every $11\frac{1}{2}$ days, about a minute in 2 years. If that is not good enough for you, let us make our imaginary clock lose only a billionth of a second every second. This clock would lose a second every $31\frac{1}{2}$ years, less than 3 seconds in most people's lifetimes!

There is an efficient and *revealing* scientific notation in which to express numbers that highlights the sizes of numbers in a very clear fashion. It is referred to as "powers of ten." This notation utilizes the fact that every number has a decimal point in it even if it is only implied. Let us first consider numbers that are bigger than 1. For whole numbers, like 530, there is a decimal point understood to be located at the end of the number. When you enter it into a calculator, it will in fact show "530." on the display. We can actually move that decimal point so that it is right after the first digit, here a 5, by keeping track of that move using a power

of ten. We would write the number as "5.3×10^2" ("5.3 times ten to the 2"), where the 2 (the "power of ten") corresponds to the number of places the decimal point was moved. Well, that may seem like a bother, but it is very useful for appreciating and comparing the sizes of numbers and also for expressing very large numbers in a simple way. For instance, 5,230,000 which is 5 million two hundred and thirty thousand would be written 5.23×10^6. It is the power of ten that tells us the *size* of this number, and therefore *the power of ten is what is of significance for us*. And what about numbers that are in the billions, requiring more than nine places? Consider 5 billion, which would be written 5,000,000,000 in the ordinary way. Writing 5×10^9 is much simpler and more revealing of the actual size of the number. We shall encounter numbers that are in the trillions (10^{12}), millions of trillions (10^{18}), etc., so this powers-of-ten notation will be extremely convenient and transparent; a million trillion (10^{18}) would be written 1,000,000,000,000,000,000 in our every day language. In fact, confronted by such a number we would naturally start counting *how many* zeros there are — *the powers of ten notation has made this count for us*. We usually don't even bother to express such large numbers in terms of the "illion" words; the power of ten tells us quickly and precisely what we want to know.

For very small numbers, the powers-of-ten notation puts a minus sign in front of the power of ten that 1 is being divided by; for instance, a millionth is written 10^{-6}, a billionth is written 10^{-9}, etc. The decimal notation for a millionth is .000001 and that for a billionth is .000000001, which are again much less transparent or efficient.

For easy reference, let us list some of the powers of ten and their word equivalents:

one hundred	—	10^2	one hundredth	—	10^{-2}
one thousand	—	10^3	one thousandth	—	10^{-3}
one million	—	10^6	one millionth	—	10^{-6}
one billion	—	10^9	one billionth	—	10^{-9}
one trillion	—	10^{12}	one trillionth	—	10^{-12}

Now we explore the basic quantum properties of nature. Our conceptual understanding of particles and waves will be greatly modified by these considerations. We will also learn some fundamental aspects of measurement and "reality." These concepts will be of great importance for us when we study the nature of particle symmetries.

Chapter 4

THE QUANTUM PROPERTIES OF NATURE — PART ONE

1. CLASSICAL PARTICLES AND WAVES

Individual properties

We have already mentioned the dictionary description of a *particle* as a massive object located at a point in space. However, this *point particle* is merely a convenient *concept*. We refer to it as a *classical* particle because, as we shall see soon, the particles we find in our quantum world have much more complicated behavior than would such an object.

A *wave* is a completely different type of phenomenon — a vibration (usually of matter) spread over a region of space. Examples of waves are all around us: ripples on the surface of water, vibrations of violin, guitar, or other musical strings, of drumheads, of air columns (in wind instruments), etc.; even sound itself is a wave. The kinds of waves we will be interested in are electromagnetic waves — vibrations of electric and magnetic disturbances themselves, that exist as distinct entities rather than mere vibrations of the particles of a medium; they display a wavelike nature under ordinary circumstances. The concept of a wave of spread out energy is referred to as a *classical* wave.

A *simple* classical wave is composed of a succession of peaks and troughs (valleys) as shown in the "snapshot" in Fig. 4.1; it extends without end in both directions. It is called a *sinusoidal wave*. It moves along at a constant speed, keeping the same shape — each part of the wave is moving at the *same* speed so the pattern moves without distortion.

The distance between successive peaks (or successive troughs) is called the *wavelength*. We describe different locations on the waveform itself as different *phases*. A peak is a completely opposite phase to a trough and different locations between peaks and troughs correspond to different phases, as well. Two waves, of the same wavelength, are *in phase* when the peaks of one overlap the peaks of the other (and hence all corresponding parts overlap). In Fig. 4.2, we see two sinusoidal waves,

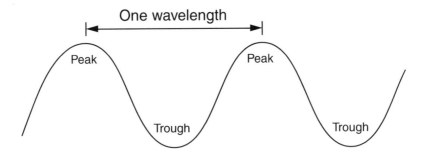

Fig. 4.1. "Snapshot" of a piece of a sinusoidal wave showing the alternating peaks and troughs. A wavelength is the distance from one peak to the next peak (or one trough to the next).

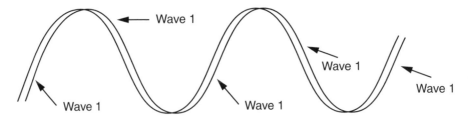

Fig. 4.2. Two waves that are slightly out of phase. A slight shift of wave 2 to the right or wave 1 to the left will make the waves overlap. After this shift is performed, the waves will be *in phase*.

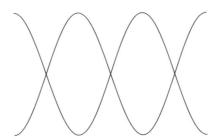

Fig. 4.3. Two waves that are completely out of phase.

with wave 1 slightly to the right of wave 2 (unmarked). If we were to move wave 2 slightly to the right or wave 1 to the left, they would overlap and would be *in phase* with one another. Sinusoidal waves are *completely out of phase* when each peak of one is at the same place as a trough of the other and vice versa, as shown in Fig. 4.3. In between these two extremes, the waves are *partly out of phase*.

Suppose that in Fig. 4.2, we continue to push wave 2 to the right past the overlapping position. It will become out of phase with wave 1 — completely out of phase when we have pushed it exactly one half of a wavelength past that position as in Fig. 4.3. When we reach a full-wavelength displacement, it will again overlap

wave 1 so that the waves are again in phase, although a different peak is now overlapping each of those in wave 1. The same will be true for a displacement of any number of complete wavelengths.

There are two different kinds of motion associated with a wave. Imagine two buoys in your favorite body of water. Suppose that a water wave is moving from one to the other. A peak of the water wave reaches the first one and that buoy moves up, then down again as a trough arrives, then up as the next peak arrives, etc. Some time later that same (first) peak reaches the second buoy, which responds similarly. The movement of the wave from one buoy to the other is the first kind of motion, and the time it takes the peak we are watching to reach the second buoy is determined by the *speed* of the wave. (In the case of electromagnetic waves, like light, the electric and magnetic disturbances travel at the speed of light.) The bobbing up and down of the buoys is the second kind of motion, which is oscillatory, i.e., repetitive in nature. The time it takes for a buoy to make one complete oscillation (up, down and then back to the original position), which is referred to as the *period*, is the time from the arrival of one peak to the arrival of the next one. In other words, it is the time it takes for the wave to travel a distance of one wavelength. The rapidity of the oscillations, called the *frequency*, is usually represented by the number of oscillations per second;* it is equal to the number of peaks passing the buoy per second. Every time the wave moves a distance equal to one wavelength, one complete wave passes our buoy causing it to undergo one complete oscillation; consequently, the frequency of a sinusoidal wave depends on its wavelength. *The shorter the wavelength* the closer together are the peaks, so that more peaks will pass per second producing *a higher frequency of oscillation.* See Fig. 4.4.†

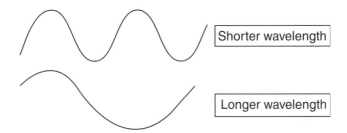

Shorter wavelength

Longer wavelength

Fig. 4.4. The upper wave has a wavelength that is one half that of the lower wave.

*The period and the frequency are inverses of one another. For example, if the frequency is 2 oscillations per second, then the period (of time) of each of those oscillations must be $\frac{1}{2}$ second (which is indeed, 1 *divided* by 2); on the other hand, 1 *divided by* $\frac{1}{2}$ (the period) is 2 — which is indeed the frequency.

†The speed of the wave is the distance covered, here one wavelength, divided by the time interval, here one period. Because the frequency is 1 *divided* by the period, the speed of the wave is also one wavelength *times* the frequency. Since wavelength *times* frequency must always equal the speed of the wave, the shorter the wavelength the higher the frequency, etc.

For some phenomena, the size of the frequency manifests itself in unexpected ways — for sound it corresponds to pitch, for light to color. When we look at a *rainbow*, as we move our eye from the red to the violet, we are moving up in frequency (red is the lowest, violet the highest frequency). Electromagnetic waves with frequencies above the violet are called ultraviolet (UV); then come x-rays and finally gamma rays. These dangerous rays are not in a rainbow, but don't look directly at the sun (especially during an eclipse) without adequate eye protection. The danger of UV and higher frequency radiation is, as we shall see, due to the fact that its *high frequency* corresponds to *high-energy* photons. Their individual energies are high enough to break the chemical bonds inside cells, including the DNA of chromosomes. Electromagnetic waves with frequencies below visible light are infrared (with frequencies just below that of red light), microwaves, radio waves, and then extra low frequency (ELF) radiation.

Electromagnetic Waves

(inspired by "The Rain in Spain" with apologies to Lerner and Lowe)

Frequency
is the key
to electromagnetic energy.
For electromagnetic energy,
frequency
is the key.

The frequency, it corresponds to color
(in the visible region).
The frequency, it corresponds to kind.
And what we mean by "kind"
is quite plain, when we say
that higher up we find
the UV
and X-ray.
And higher come from nuclear decay
(called gamma ray).

For frequency, it tells *the photon secret*
(a quantum feature).
It is a measure of each energy.
More energetic ones,
'tis quite plain, as we see,
can break your DNA bonds
without strain;
it's easy!
The danger 's as apparent as can be
(we all agree).

Frequency
is the key
to electromagnetic energy.
For electromagnetic energy,
frequency
is the key.

Combined effects

The combined effects of more than one classical wave have a very different nature than do those of combinations of classical particles. When two such waves arrive at a point in space, the resulting pulsation depends on the relative phase of the waves. For a wave is an oscillatory disturbance, alternately pushing the same point in a medium first one way, then the other, then the first way again, etc. We have already seen this happening as we watched a water wave move by our buoy — as the crest of the wave went by the buoy bobbed up, and as the trough (or depression) reached it the buoy sank down. When two waves (of the same size) arrive simultaneously at the buoy, it will respond to the combination of the forces exerted by the waves.

If they both arrive in a pushing-up phase the buoy will be pushed up twice as far, similarly if they are both pushing down. However, if the two waves arrive completely out of phase with one another, one pushing up and the other pushing down with the same amount of force, the net result will be no motion of the buoy at all. The buoy will not acquire any energy from the waves. These effects are referred to as interference; the cooperative effects of waves in phase are referred to as *constructive interference* (yielding a bigger wave), and the canceling effects of waves out of phase as *destructive interference*. When the waves *completely* cancel at a point, we call it *total destructive interference*.

On the other hand, when two classical particles impinge on a particular point in space, for instance firing two bullets at our buoy, the resulting energy deposited there is the sum of the energies of the individual bullets. (Bullets behave very much like classical particles in these kinds of circumstances.) Even if they come from opposite directions, so that they impart no net momentum to the buoy, the buoy will acquire the sum of the energies, which the bullets lose when they stop. (The buoy's energy will be increased by the amount of heat generated by the friction of the bullets in the buoy's material.) *Classical particles do not experience interference effects.*

<div align="center">

Two can be the same as None

When a classical wave met a classical wave as they traveled along there in space.
Said the classical wave to that classical wave, "Oh, I long for a friendly embrace."
Those two classical waves overlapped at that spot, 'twas observed then by all who were near.
But alas, in that region of overlap, *no* disturbance did ever appear.
And I'm told, by a witness right there on the scene, that the first said "But, wait there's a catch!
For you see we're completely out 'f phase! *You* must now shift your phase so we match."
But the second one was obstinate, and the first one was stubborn too.
So destructive interference won; there's no record, there's not a clue,
that one day, at that location site, a long visit was made by those two.

</div>

2. QUANTUM OBJECTS AND INDETERMINACY

Particles and waves

The particles we find in nature have much more complicated behavior than the classical point particles of our intuition described above. In fact, they display a wavelike nature under certain circumstances. Likewise, the *actual* electromagnetic waves we find in nature have much more complicated behavior than do classical waves. In fact, there are circumstances under which an electromagnetic wave behaves as if it is made up of our favorite massless particles, namely photons. Thus, we find a *wave-particle duality* in the behavior of all of the constituents of nature — under certain circumstances they behave like classical particles, under others like classical waves.

A particularly *illuminating* (pun intended) class of experiments displaying these features is one in which we have a vertical barrier in the path of a horizontal wave

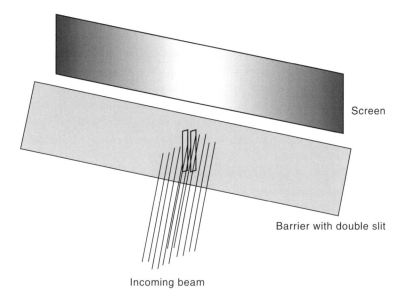

Fig. 4.5. A two-slit experiment, showing the pattern that would be made on the screen by a **beam of classical particles**. The pattern on the screen is merely the sum of the patterns made by the particles that went through the right slit and those that went through the left slit. The paths of most particles going through a slit are only slightly deviated from their original directions by tiny interactions with the slit edges; hence, most particles will arrive near the center of the screen.

or particle beam. Consider a barrier that has two side-by-side parallel slits cut vertically into it as shown in Fig.4.5. What will we see on a screen beyond the barrier? For a classical particle beam, some particles would go through the left slit, some through the right. We would see a pattern of particle intensities which is the *sum* of the intensity pattern we would have had from the left slit alone and the one we would have had from the right slit alone, as shown in Fig. 4.5.

Not so for a classical wave! A different piece of the wave emerges from each slit, each behaving like a separate wave. Those emitted at the same time are always in phase with one another; however, the phase of the waves being emitted oscillates (up, down, up, etc.), as time passes. Let us consider the center of the screen, which is equidistant from the two slits. At any time, it is receiving a wave from each slit. Since each traveled the same distance (from slit to screen), they were emitted at the same time and so are in phase. As time elapses, the phase of these arriving waves oscillates in time; however, the frequency of oscillation is about a thousand-trillion (10^{15}) oscillations per second, so rapid that we do not detect the on-off-on nature. Instead, the illumination appears bright; consequently, there will be a bright line (similar to the shape of the slits) at the center of the screen, due to constructive interference. However, consider a place slightly to the right of the center of the screen. Again we consider the waves arriving at the same time. What will their

relative phases be? The wave that came from the left slit traveled a greater distance and so had to have been emitted earlier than that from right slit. Therefore it is slightly out of phase with the one from the right and the screen will be less bright at that place. As we go further to the right on the screen, the time difference of emission and consequently the phase difference of the two arriving waves is larger and larger. Eventually, we will reach a place on the screen where the waves arriving are completely out of phase resulting in total destructive interference. That place will be *totally dark*. Further yet to the right will be a position where the waves are again in phase. That position will be *bright*. Moving beyond that position we will find the phase differences continually changing from in phase to out of phase to in phase, etc. (A similar pattern is produced on the left side of center.) Thus we will see *a series of alternating bright and dark lines* on our screen due to the interference properties of waves, as shown in Fig. 4.6.

Two-slit interference patterns of electromagnetic waves (light) were seen as long ago as 1801. We don't see such patterns in our everyday experience for a couple of reasons: The slits must be very close together (a small fraction of an inch apart for light) for the pattern to be evident; and ordinary bulbs emit *short pieces* of waves with *random* phases, so that phase correlations between any two are constantly changing.

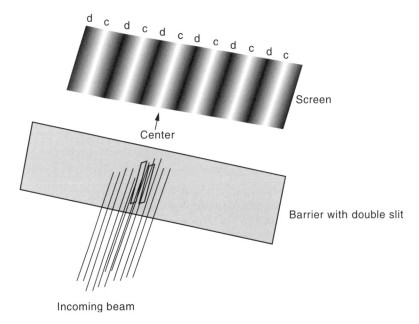

Fig. 4.6. A two-slit interference experiment, showing the pattern that would be made on the screen by *a **coherent** wave*. The wave is moving from barrier to screen. On the screen, c marks regions where constructive interference produces bright lines, d marks those where destructive interference yields dark lines.

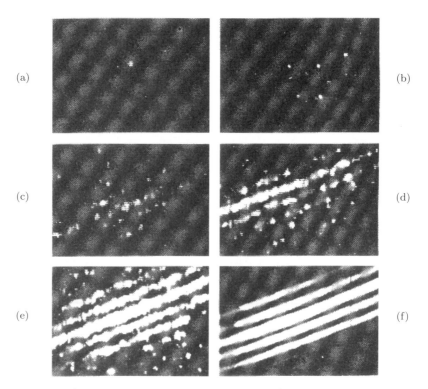

Fig. 4.7. Patterns (observed on a TV monitor) of electrons that have gone through a barrier containing two slits. Each spot corresponds to one electron hitting the TV screen, and each successive picture corresponds to a larger number of electrons. Notice how the individual spots build into an interference pattern. (From Merli, Missiroli and Pozzi, *Am. J. Phys.* **44**, 306 (1976)).

When we perform a similar two-slit experiment with a beam of *electrons*, we can see one electron at a time arriving at the screen, making a single mark just as we would expect for a classical particle. However, after many electrons have left their marks on the screen, we see a pattern *identical* to the two-slit interference pattern of classical waves, with dark lines (where no electrons hit at all) and bright lines (where many electrons hit), as shown in Fig. 4.7.

So the electron *distribution* displays wave-like behavior. In contrast, a *classical* wave would have produced such a spread-out pattern at once. Even if the beam of electrons is so weak that only one electron comes through the system at a time, we will see an interference pattern building up after a large number have come through. Each separate electron of the beam has absolutely no interaction with any of the others that precede or follow it. Each goes to only one place, yet after a while an interference pattern is apparent. So individual electrons are particle-like but the particles follow a distribution pattern that is wave-like. (Exactly the same interference phenomena occur for neutrons, protons and all other elementary particles.) Somehow, there must be a wavelike property associated with the electron

that senses both slits and experiences interference; its mathematical expression is called the *wavefunction*. However, the effect of the wavefunction shows up only in the cumulative pattern formed by many individual electrons. It is analogous to tossing a coin. The probability of getting heads is $\frac{1}{2}$ (for an unloaded coin), that for tails is $\frac{1}{2}$ as well; but a single toss will not indicate this at all, since we get one or the other. Even ten tosses have a good chance of not producing five and five. However, a million tosses will produce very close to half each, thus revealing the underlying probability for each (which we call the *probability distribution*). Likewise, the cumulative pattern that the electrons make on the screen reveals the wavelike *probability distribution*. It is not that successive electrons are interacting with one another; we can send in only one a second (a minute, an hour, a millennium), so that each one is alone (the previous electrons having already made their marks) and the wave pattern will still emerge after a large number of hits. We conclude that the wavefunction represents a *probability* distribution that *each* individual electron must obey.

But we could ask: if each electron went through one slit or the other, why don't we see instead a pattern that is the sum of what we would get from each of the slits, as shown in Fig. 4.5 — a pattern with no dark lines? To investigate this further, we could set up a very tiny detector to tell us which slit the electron went through; a simplified sketch showing how that may be accomplished is shown in Fig. 4.8.

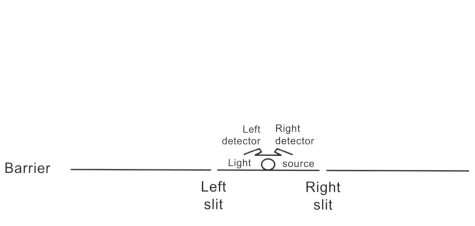

Fig. 4.8. Top view of a two-slit scattering experiment in which a slit detector is located between the slits. The slit detector consists of a light source to send out photons, one of which will scatter from the electron as it emerges from a slit, and two electric eyes to detect the scattered photon. If the electron emerges from the left slit then the left detector will flash, etc.

We could thus double-check the distributions of those that went through each slit to see whether their sum *could* give the interference pattern on the screen. However, when this is done we find that whenever the slit detector is turned on, the interference pattern disappears and the distribution becomes just what we would expect for a *classical particle* beam. As soon as we turn the detector off, or make it too weak to determine which slit the particle went through, the interference pattern (of Fig. 4.6) returns. Thus, the detector causes an enormous change in the probabilistic wavefunction of the electron (hence in the behavior of the electrons) even if its interaction with the electron is quite weak, as long as it can determine which slit each electron went through.

In this particular case, the effect of the measurement can be seen to be a consequence of the quantum nature of light. In order to determine the location of an electron as it goes through the barrier, we must use light whose wavelength is about as small as, or smaller than, the distance between the slits. Much longer wavelengths are not sensitive to that distance. However, we shall soon see that *short wavelengths* in the wavefunction *correspond to high momenta* for the corresponding quantum particles. So the photons that determine which slit an electron has gone through must have momenta *greater than or equal to* the momentum corresponding to the longest useful wavelength. When one of these photons scatters from the electron, the electron obtains some momentum perpendicular to the original direction of the electron beam, in other words it is kicked sidewise; different electrons obtain different (random) amounts. Consequently, the electron distribution on the screen is more uniform and the dark lines are gone. This argument has mixed together our classical ideas about optical phenomena, namely needing short wavelengths to see short distances, with the quantum connection between wavelength and particle momentum. We shall soon see that quantum theory is counter-intuitive, and so physicists *often* find it useful to employ classical models and pictures to visualize what is happening followed by the almost magical insertion of quantum properties and restrictions where needed. The word *semi-classical* is used to describe these kinds of considerations. Usually, their predictions are qualitatively valid; however, we shall see (in the next chapter) that there *are* experiments in which the influences of the detectors on the beam cannot be understood in a semi-classical way. Even in those situations, the detectors still profoundly affect the results.

So, in the process of obtaining an understanding of the *actual* particles of nature, we have found that the world behaves in very strange ways. That behavior is explained by the quantum theory of particles called "quantum mechanics," which was developed in the first quarter of the twentieth century. By this time, at the beginning of the twenty-first century, the odd, seemingly bizarre predictions of quantum mechanics have been tested and verified so many times that no scientist doubts the quantum nature of matter.

Indeterminacy

The probabilistic property of quantum mechanics contains as a consequence an indeterminacy that is counter-intuitive. Before the twentieth century, an underlying assumption of all of science was the belief that the conditions and properties of a particle or system at any given time (including anything with which it will interact) completely determine what will happen to it next. An absolute determinacy lay behind every investigation and theoretical conjecture. However, the experimental discoveries in the beginning of the twentieth century *forced* scientists to abandon that revered principle for a quantum theory that is intrinsically indeterministic. In quantum mechanics, the basic object whose evolution in time *is* deterministic is the wavefunction, which is not directly observable. Instead, it serves as the basis for predicting the *relative* probability of *different* outcomes in the future.

A familiar example of this probabilistic, indeterministic behavior of nature is radioactivity. In a chunk of radioactive material containing identical nuclei that were all produced at the same time, each decaying nucleus emits one or more particles as it decays into another type of nucleus. These decays are often characterized by the *half-life*, which is the time it takes for one half of the nuclei to have decayed. These nuclei, all in the same state to begin with, do not behave in an identical manner. Half decay during that time interval and half do not. Therefore, a particular nucleus in our chunk of radioactive material, which is in a state identical to that of any of the other nuclei, may or may not have decayed during that time; the probability that it has decayed during that time is $\frac{1}{2}$.* This indeterminacy is inherent in our quantum world.

The indeterminacy of quantum mechanics led several leading scientists to reject it as an acceptable theory of the physical world. Einstein is quoted as having said: "God does not play dice with the universe!" However, every prediction of quantum mechanics has been verified so that the only apparent way out of accepting a probabilistic theory of the world is to say that the descriptions of particles, nuclei, etc., are not complete. This approach would claim that the nuclei decaying at any particular time after their creation are actually different from those that decay at another time. In other words, there are hidden properties, technically known as hidden *variables*, not contained in our quantum theory, that distinguish *seemingly* identical systems. However, we shall soon encounter the experimental evidence that rules out meaningful hidden-variable theories.

The wavefunction, whose time evolution is deterministic in quantum mechanics, behaves like a wave, but it is *a probability wave* rather than a wave of energy moving in a medium. It determines the *probability distributions* for various types of experiments, like the two-slit experiment described above, rather than the definitive behavior of any individual particle. Each individual particle makes *a spot* on the

*In case you are wondering why we don't try to measure a "whole-life," remember that when only a few are left they will no longer accurately characterize the probability distribution.

screen, and only after many have hit will a wave pattern be formed by the aggregate of the spots. On the other hand, if a classical wave goes through a two-slit barrier, the interference pattern will appear as soon as the first part of the wave reaches the screen. Thus, although a beam of quantum particles does have many of the features of an ordinary classical wave, those wave properties (of the wavefunction) show up in the aggregate behavior of the particles. We have already seen one such example — the two-slit interference pattern produced by a beam of electrons. Actually, by the end of the twentieth century, several analogous experiments had been done with very dim light and detectors sensitive enough to detect even one photon. Researchers were able to watch spot after spot light up until an overall pattern could be discerned — the interference pattern that had been seen in 1801!

<u>Quantum Particles</u>

We are the regions of the screen, on which the pattern fell,
and there is such a strange, mysterious saga we must tell.
Electrons, one by one, were fired to where we all did sit.
A barrier was in the way with only one thin slit.

As time went by, we *each* were hit by some of them; 'tis true
that those in-line with th' slit had more, as you'd expect them to.
But all of us were hit by some and none was left deprived,
There 's nothing so unusual about where they arrived.

But when a second slit was opened what a weird effect.
For suddenly electrons then began to deselect
a half of us, who they avoided hitting, with great care.
The other half were getting many more than their fair share!

Well, how could that possibly be? Tell me!
Well, how could that possibly be?

I sat there like a fool and smiled, but none would come my way.
Yes, half of us were left alone to idle all that day.
And those between us were so busy, with their extra guests.
We could not help but envy them, bewildered by our *rest.*

Well, how could that possibly be? Tell me!
Well, how could that possibly be?

They say the explanation is: these "particles" behave
according to prob'bility that's very like a wave.
Their *distribution* brings the interference into view,
destructive in our places, we get none (instead 'f each two).

Well, how could that possibly be? Tell me!
Well, how could that possibly be?

And something weirder yet occurred; that *they've* explained I 'dmit.
When 'lectric eyes were put in place, recording at each slit,
the ones of us who'd been denied now *did* receive each two
of the recorded particles, we thought that we were due.

Well, how could that possibly be? Tell me!
Well, how could that possibly be?

Wavefunction is the crucial clue; it is the secret key.
A measurement of "part'cles' " paths, destroys most certainly
the interference, since the state's wavefunction has been *changed.*
The probability is consequently rearranged.

Well, how could that possibly be? Tell me!
Well, how could that possibly be?

Now we continue our struggle with the quantum enigmas we have just met and proceed to wrestle with their arcane version of "reality."

The wavefunction

(with apologies to lovers of "Frère Jacques")
What is waving, what is waving?
I don't know, I don't know!
Probability is, probability is.
That is so, yes that's so.

Chapter 5

THE QUANTUM PROPERTIES OF NATURE — PART TWO

1. QUANTUM PARTICLES AND UNCERTAINTIES

Particle attributes of the wavefunction

The first hint of the particle aspects of electromagnetic waves was found by Max Planck at the beginning of the twentieth century, although he did not recognize it as such at the time. (Albert Einstein was the one who recognized the significance of Planck's discovery.) In the course of his work, he discovered an extremely small constant of nature that we refer to as *Planck's constant*, which is denoted by the letter h. In Standard International (SI) units this number is roughly 6 *divided by* the enormous number 10^{34} (1 followed by 34 zeros) and is written 6×10^{-34}. It is about 3/5 of a billionth of a trillionth of a trillionth, by far the smallest physical number we know.

For a simple sinusoidal wavefunction, it has been found that the momentum of the quantum particle described by this wave is Planck's constant *divided by* the wavelength*; thus, *a short wavelength corresponds to a high momentum*. The particle's energy is Planck's constant *times* the frequency, so that a *high frequency corresponds to a high energy*. Since high momentum implies high energy, these quantum-particle relationships are consistent with the fact that short wavelengths correspond to high frequencies, as we saw in the previous chapter.

For actual particles, which do not extend over all of space, the wavefunction cannot be a simple sinusoidal wave (which does). Instead the wavefunction may be understood as a sum of sinusoidal waves of different wavelengths that destructively interfere outside the region occupied by the particle. In Fig. 5.1 we see how even two waves of slightly different wavelengths can be in phase in one region but out of phase in others.

Actually, the two waves in Fig. 5.1 come back into phase again further out in both directions. For complete destructive interference *everywhere* outside the

*commonly referred to as "the DeBroglie wavelength" of the particle.

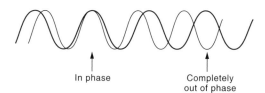

<center>↑
In phase</center> <center>↑
Completely
out of phase</center>

Fig. 5.1. A comparison of two waves of different wavelength — the darker one has the longer wavelength. Notice how they go increasingly out of phase the further away from the in-phase region we look.

Fig. 5.2. A wavepacket, that is essentially zero outside the region shown.

central region, we need an infinite number of waves, one for each wavelength in a (continuum) spread of values. Such sums are often referred to as *wavepackets* and they look roughly like the waveform shown in Fig. 5.2.

Let us explore this notion of a sum of sinusoidal waves and its consequences:

Uncertainty relations

As we have already discussed, in a quantum mechanical world, measurements are not as simple as our intuition expects. We have already seen that a measurement of the location of an electron as it goes through a two-slit barrier has a dramatic effect on the subsequent behavior of that electron. For example, consider a place on the screen that corresponds to a dark line when the slit detector is off, i.e., when no sidewise position measurement is made at the barrier. That place is dark because no electron goes to that position on the screen. With the position detector on, however, the interference pattern disappears so that that position is no longer dark — some electrons *do* arrive at that place on the screen. Those electrons thus have completely different behavior than they would have had were the detector left off. The detector's measurement of their sidewise position along the barrier evidently changed the sidewise speed or momentum with which they subsequently traveled. In fact, some electrons have their momenta changed by large amounts, some by small amounts, and some not at all; individual changes are uncontrollable and unpredictable. It is very similar to trying to confine a water wave to a small region; it becomes very choppy. The smaller the region of space we confine the electron to, the more "choppy" is the momentum distribution of its wavefunction, which gives the relative probabilities for different possible values of the electron's momentum. Thus, *the smaller we make the position uncertainty of the particle, the greater is its momentum uncertainty.* In order for the particles' sidewise momenta to remain zero,

we would need to remove the barrier altogether, which amounts to complete uncertainty of their sidewise positions. This and many other more detailed experiments confirm this quantum-correlation between position uncertainty and momentum uncertainty. This relationship is referred to as *the Heisenberg uncertainty principle*. It should be noted that this quantum-correlation of position and momentum measurements is usually not observable for macroscopic-size objects. This is so because the quantum-correlation of the uncertainties depends on Planck's constant, which is a very tiny number; the accuracy of usual measurements on macroscopic systems is not good enough to get the uncertainties down to that level. It is only when we can measure properties of small numbers of particles, or particles that are all in exactly the same quantum state, that quantum mechanical uncertainty relations become evident.

In summary, we can say that the uncertainty principle arises in a natural way in quantum mechanics, since particles are described by wavefunctions. In the case of a sinusoidal wave, *the momentum of the particle is determined by the wavelength* of the wavefunction. In the three-dimensional space of our universe, a sinusoidal wavefunction extends over all of space in every direction; consequently, it does not describe a particle at all! In order to describe a particle *localized* in some region of space, the wavefunction must have waves of *different* wavelengths, which can destructively interfere outside that region — *the smaller the region, the larger the required spread of wavelengths*. But a spread in wavelengths corresponds to a spread in the momentum distribution contained in the wavefunction. Thus, the uncertainty principle results from: (1) the property possessed by waves, that a localized wave must contain sinusoidal waves over a *spread* of wavelengths — the smaller the region of localization, the wider the spread; and (2) the quantum mechanical connection between wavelength and momentum.

Uncertainty

And built into nature, we can see
uncertainty, uncertainty.
It's hard believing we're not free
of uncertainty, uncertainty.
It doesn't impugn our ability,
this uncertainty, uncertainty,
but nature's teaching humility
with uncertainty, uncertainty.

You can measure the position.
You can measure the position.
You can measure the position,
with the very best precision!

You can measure the momentum.
You can measure the momentum.
You can measure the momentum,
So precise, with great invention!

But not both!

Interchange — Independent or Quantum-conjugate

Let us now imagine a chef at work. She is preparing several dishes, but we only need to look at a couple of them to learn the concepts we are after. She is preparing a salad; see how she puts the dressing on top and then tosses the salad. Suppose she had *interchanged* the order of those operations, tossing the salad first and then putting the dressing on top. The result would be completely different! I don't know about you, but I prefer my dressing mixed in! The two operations we have witnessed, putting dressing on top and tossing the salad, are obviously not independent of one another, since the order in which they are performed is crucial. We can say that in some sense these two operations are *related*.

Now watch as the chef butters the croissants. Does it matter that she put them on the platter before buttering? What if she would have buttered them on the table and then put them on the platter; if we had looked away until they were sitting there beautifully buttered, could we have detected which was done first? Before you answer, I must caution you that these are croissants that do not make crumbs when being buttered and she is an *expert* butterer. No, the interchange of these two operations, putting them on the platter and buttering would have no effect. These two operations are completely *independent* of one another.

Let us now consider an extremely simple experimental setup. Suppose we have only one narrow slit in a barrier. A beam of quantum particles is incident on it and a screen is placed several feet away on the other side. The quantum particles' distribution as they emerge from the slit will have a wavelike probability that behaves like a classical wave (including interference). There will be a pattern with an unexpectedly wide bright region in the center of the screen — much wider than the width of the slit itself; it is caused by the interference of the (probability) waves "coming" from different locations across the width of the slit. It is called a diffraction pattern because it looks as if the beam were *diffracted* or bent outward by the edges of the slit. Figure 5.3 shows such a diffraction pattern.

If we measured the sidewise momenta of the beam particles *before* they reached the slit, we would find them to be zero or very very small (zero within experimental error); after all, it is a beam of particles moving *toward* the barrier. However, after they go through the open slit, some of the particles acquire sidewise momenta, as is evidenced by their arrival at widely dispersed regions on the screen. (In fact, their position on that screen can be used as a measure of their sidewise momenta.) So a measurement of their sidewise momenta, *after* their sidewise position was determined to be within the slit opening (since they did go through it), gives a completely different result than would be obtained by the measurement of zero for their sidewise momenta made *before* they reached the slit. The shorthand expression of this relationship between position and momentum (in the same direction) is to say that they are *a quantum-conjugate pair of measurements — the order of their*

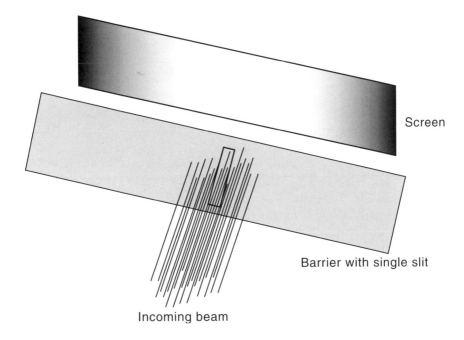

Screen

Barrier with single slit

Incoming beam

Fig. 5.3. A single-slit diffraction pattern.

*measurements determines the result.** More generally, the effect that a measurement has on the subsequent behavior of a particle implies that successive measurements of quantum-conjugate properties may give markedly different results, depending on the order in which the measurements are made. Furthermore, the mathematics shows that quantum-conjugate pairs of observables, like position and momentum, must obey an uncertainty relation — the smaller the uncertainty in the measurement of one of them, the wider the spread of the other.

<div align="center">

Quantum-conjugate Measurements

</div>

<div align="center">

A young physicist claimed he'd observed
two quantities (that are conserved).
That that clashed with the theory,
Made the referee leery.
Was the paper's rejection deserved?

</div>

To indicate that the interchange of two operations is undetectable, we say that they *commute* (with one another); this has nothing to do with traveling — a completely different meaning of the word. If two operations are quantum-conjugate then they do not commute. Therefore, we can say that the position and momentum

*Technically, the designation "quantum-conjugate" requires more than the fact that the result of successive measurements depends on order, but the distinction is of no importance here. The pair position & momentum do indeed form a quantum-conjugate pair.

operators do not commute. We shall encounter other sets of operators that do not commute as we approach *gauge theories* in our travels.

<div align="center">

Quantum-conjugate

do they *commute?*
they do *commute,*
NOT!

</div>

An uncertainty relation exists also for energy and lifetime. The quantum mechanical argument is somewhat analogous to the one we just looked at for position and momentum. The energy associated with a particle is directly related to the frequency of the (probability) wave describing that particle. However, a (sinusoidal) wave of one frequency oscillates forever. For a wave to die out with time, it must have sinusoidal waves of different frequencies in it to interfere with one another. The shorter the time of its existence (referred to as its *lifetime*), the wider the spread of frequencies that must be present to accomplish the cancellation. Thus lifetime and frequency have an uncertainty relation connecting them, in the same way that position and momentum do. In this quantum world of ours, however, the frequency of the wavefunction is directly related to the energy of the corresponding particles. For particles at rest, their energy is characterized by their (rest) masses (recall $E = mc^2$) so that we have *an uncertainty relation between lifetime and mass!* Among its consequences is the fact that particles that have finite lifetimes will not have definite masses. Instead, if we measure the individual masses of a collection of such particles, we will find a *distribution* of values; they will not all have the same mass! The distribution will be large in the center, i.e., most such particles will have masses close to that central value; however, some may have masses quite different from that value. The shorter the lifetime, the wider the mass spread.

Quantum sizes of particles

The wavefunction of a particle is not located solely at a point; it takes up a finite region of space, as all waves, including wavepackets, do. Therefore, a particle sitting at a point in space has an effective size, which is not infinitesimal. The lower the mass of the particle, the larger its effective size.

<div align="center">

Quantum Rebel

A particle 's spread like a wave
that *no* point can ever enslave.
So *quantum* in manner, it taunts,
as classical notions it flaunts.
No matter how we may enjoin 't,
the point is it can't be a point!

</div>

2. QUANTIZATION

Quantized energy levels

Our first inkling of the quantized structure of atoms resulted from observations of light emerging from them. Many different experiments showed that when atoms are heated, bombarded, or jolted with electricity, they emit certain definite colors of light. Each type of atom has its own spectrum of colors which, unlike the colors of the rainbow, are isolated (discrete) and distinct rather than continuous gradations of color. Different colors correspond to different frequencies, so that the frequencies of the emerging light do not form a continuum but instead are quantized. Studies of light emerging from hot objects and from atoms eventually led to the development of quantum mechanics, one quarter of the way into the twentieth century. The detailed predictions of quantum mechanics have been checked very many times, and applications of that theory have led to a large number of techno-logical developments as well. These include the transistor and all the subsequent electronic technology based on it. Quantum mechanics has revealed the structure and behavior of molecules, knowledge that has led to many medical achievements including the determination of the causes of diseases, the invention of new medi-cations, the creation of tracers and other diagnostic tools, all the way to genetics (and genetic treatments). A large number of developments in chemistry and its industrial applications have also been made.

We shall soon see that the *wavelike* character of the particles of nature leads to the quantization of their energies in matter and ultimately to the stability of matter. In order to explore this matter (definitely a pun), let us first consider a familiar classical wave: the wave that we set up when we pluck a string of a guitar or violin (or *your* favorite stringed instrument). We hear a definite tone. In addition, there are overtones of it that give the instrument its distinctive sound — a different relative-loudness distribution of the overtones for different instruments. (That is why a note plucked on the violin sounds different from the same one plucked on the guitar.) The point is that there are only certain definite pitches excited when we pluck the string; referring to the frequency associated with the perceived tone as f, ideally we also excite some amount of the frequencies *2 times f* (the first overtone), *3 times f*, *4 times f*, etc. For actual strings, the frequencies of the overtones may be slightly different, but nevertheless their frequencies are *quantized*. They are restricted to certain definite isolated (discrete) values; they do not have a continuum of possible values.

When we are dealing with the *wavefunctions* that describe particles, the fre-quencies correspond to their allowed energies. For electrons in an atom (or a solid, etc.), we have a situation analogous to a string of an instrument and the result is that only certain allowed energy states exist for the electrons; *their energies are quantized*.

<u>Quantization in My House</u>

A half a step you can't alight;
you cannot walk down less.
And *more* may still not be just right.
Exactly one? Oh, yes!
Or two or three or four or five,
or six or seven; more?
Well reaching twelve you will arrive
right at the basement floor!

Quantized spin

The quantization that results because of the wave nature of particles is not restricted
to energies but also appears in the intrinsic angular momenta of particles known as
spin, as we have already discussed in Chapter 2. Furthermore, the orientation of the
spin is quantized as well. In particular, particles that have spin $\frac{1}{2}$, like electrons (and
all the other fundamental matter particles), have only two possible orientations of
their spin — they can be rotating clockwise (called spin down) or counter-clockwise
(called spin up) about any direction chosen for the measurement. Thus, for a spin
$\frac{1}{2}$ particle, we have two possible states for every state that it would have if it had no
spin. *Massless* gauge particles like the photon and its cousins also have two possible
orientations of their spin.

The Pauli (pronounced Powli) Exclusion Principle

Quantum theory also teaches us that fermions — e.g. electrons and, in fact, all
of the fundamental matter particles (leptons and quarks) — have a solitary na-
ture. In fact, they obey a strict *exclusion principle* — once a fermion is in a
particular physical state, all other *identical* fermions are excluded from it. Thus,
this Pauli Exclusion Principle shows that no two identical fermions can be in the
same state.

<u>KEEP OUT</u>

You cannot come in here, I say,
although there is no sign.
I need my state; I will not sway.
This state is solely mine!

And if you think I'm being mean
and selfish as can be,
I ask, "Haven't you *ever* seen,
the FERMION DECREE?"

THOU SHALT NOT SHARE YOUR STATE 's the rule;
we may not share at all!
Now don't get angry, please stay *cool*;
but just respect my "wall."

A classical cataclysm

In classical electrodynamics, it is shown that a *charged* particle radiates electromagnetic energy as its momentum changes (direction and/or size). Charged particles moving around the oval beam paths in accelerators display this property very dramatically, producing radiation (called synchrotron radiation) that has been put to medical and other important uses. Similarly, in a classical world, the electrons circling the nucleus in an atom would emit a steady stream of electromagnetic waves. This continual loss of energy would cause them to spiral down into the nucleus. In this manner, all classical atoms would collapse shortly after they are made! Of course, the very fact that we *exist* shows that such cataclysmic collapses do not occur. After all, we and everything around us are *made* of (uncollapsed) atoms. In other words, *the fact that we exist to create and consider the classical theory of nature precludes its validity.*

The classical cataclysm averted and chemistry explained

It is the fact that the allowed energies of electrons in atoms are quantized, along with the Pauli Exclusion Principle, that explains the stability of atoms — their violent collapse is averted. Envision a skyscraper shaped like an ice cream cone with the bottom broken off, containing no stairs and no basement, as shown in Fig. 5.4.

The floors have corridors that circle around the cone back onto themselves. The ground floor has one corridor and successive floors have more and more (concentric) corridors. The different floors correspond to different allowed energies of the electrons in an atom; they are referred to as *energy levels*. Like the floors of any skyscraper, they are not a continuous set, but are spaced instead — the allowed energy levels are quantized. An electron on a particular floor will be spread out throughout a particular corridor like a wave (or like cotton candy), but no electron can exist between floors. An electron can make a transition to a lower floor by

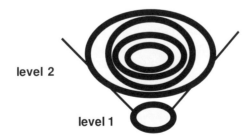

level 2

level 1

Fig. 5.4. A *simplified* representation of the energy levels, and the corresponding states, of an atom (showing the lowest two levels). As we go up this 'energy-level skyscraper,' there are more and more 'corridors' (shown as **black** ovals) representing the possible states of the electron. Each oval represents two states, one corresponding to spin up and the other to spin down.

coalescing long enough to enter the elevator. (This "coalescing" to enter an elevator is only meant metaphorically. The actual process cannot be followed that closely.) In this *particle-like* condition, the electron will give off energy in the form of a photon and wind up on a lower floor (a state of lower energy) where it will again spread out in one corridor. To go up, the electron must obtain the extra energy required to reach a higher floor. This may be accomplished by having a photon close enough for the electron to absorb it. It is significant for the stability of atoms that there is a ground floor, below which the electron cannot go — the atom will not collapse! Furthermore, only two electrons (spin up and spin down) can exist at the same time in any corridor, due to the exclusion principle.

The number of electrons in a neutral atom equals the number of protons in the nucleus, for the charges of the atom must add to zero. In an unexcited atom, the electrons *fill* the lower levels (floors), the more electrons, the higher the top of the *filled* states. Different elements (i.e., atoms) have different numbers of protons and that same number of electrons. Consequently, the electrons in one element fill to a different level, or a different corridor on the same level, from those in another element. All of the chemical properties of different elements result from this because chemical interactions and also the behavior of bulk material depend on interactions of the outer (highest energy) electrons of one atom with those of another. For example, hydrogen, with one electron behaves very differently in chemical reactions than does oxygen with eight, just because the other seven electrons of oxygen must all be in *different* states. A successful explanation of chemical behavior, even of organic molecules (that make up living organisms), has come from quantum mechanics.

In order that our skyscraper analogy be more faithful or realistic, we must imagine that the cone-skyscraper itself is invisible — we "see" only the electrons. Furthermore, the skyscraper would have to have an infinite number of floors, which are far apart near the bottom but are closer and closer together as we approach the roof, even though the roof is only a finite distance above the ground — impossible for an actual skyscraper, even an invisible one, but ordinary for an atom. Yes, there are, indeed, an infinite number of bound states bunched up like that. (In practice, any electrons near the top, i.e., that are very loosely bound, are so easily pulled away from the atom that they behave as if they are free.) Looking down from a helicopter, we see a two-dimensional pattern that looks somewhat like the electron distribution in an atom. (This metaphorical skyscraper is not a completely accurate representation of an atom. The wavefunctions are actually spread beyond their "corridors," and the atom is a more complicated *three*-dimensional object.)

3. QUANTUM FIELDS

As we have already seen, when electromagnetic waves (light, etc.) are sent through a two-slit barrier, they produce an interference pattern looking like that expected for a *classical wave*. However, when we use *extremely dim* light and a screen made

of small sensitive detectors, we find that the light does not spread out across the screen as a classical wave would. Instead, one tiny detector at a time receives light, just as for electrons. Thus, we verify that light and all electromagnetic waves are made of particles — photons. It is apparent, from the *interference* pattern (made of spots) that builds as more and more photons reach the screen, that the individual photons must be conforming to a probability distribution much as the electrons did in a similar interference experiment. Photons are *quantum particles* whose wavelike characteristics have been apparent to us for about two centuries; however, it wasn't until the first part of the twentieth century that we realized that electromagnetic "waves" were actually composed of photons. Even fairly dim light corresponds to so many billions of photons that we cannot distinguish the individual flashes; that is why this point-like behavior of light was not seen before the development of very sensitive detectors.

The first evidence of the inadequacy of classical electromagnetic theory to account for the behavior of electromagnetic waves was found in the study of electromagnetic radiation from hot objects (technically called "black-body radiation"). At the beginning of the twentieth century, Max Planck found that the relative intensities (brightnesses) of the different frequencies contained in blackbody radiation could be understood only if that electromagnetic radiation were being emitted in *quantized* amounts. After subsequent elaboration by Einstein and others, over a period of about fifteen years, it was understood that electromagnetic waves were actually made of photons; the quantized amounts of radiation were the energies of the individual photons. Each photon has an energy equal to the *product* of Planck's minute constant (h) *with* the frequency (f) of the wave associated with that photon. Consequently, the electromagnetic energy of frequency f in an electromagnetic "wave" must be the *product* of the number of photons *times* the tiny energy of each one (hf). Thus the energy of an electromagnetic wave of frequency f is quantized in units of hf (just as prices in the supermarket are quantized in units of pennies). Each photon of frequency f carries the same (tiny) amount of energy, whether the light is dim or bright — *the more intense* the radiation *the more photons* present. Since Planck's constant h is an incredibly small number, it is understandable that this quantization was not noticed earlier.

Let us now enlarge our electromagnetic vocabulary: An electromagnetic wave has a definite strength and direction for the electric disturbance and a definite strength and direction for the magnetic disturbance *at each point in space.* The situation is somewhat analogous to the lawn in front of my house. At each point on the lawn, there is a blade of grass with a definite height pointing in a definite direction. (Those places that are bare have a blade of height zero.) The height corresponds, in this analogy, to the strength of the electric or magnetic disturbance under consideration and the direction corresponds to its direction. This analogy to a *field* of grass leads to the characterization of the electric disturbances as an electric *field*, and the magnetic disturbance as a magnetic *field*; the wave, containing

both, is referred to as an electromagnetic *field*. (We also use "electromagnetic field" generically to refer to either one or any possible combination of electric and magnetic fields.) Although my lawn is a two-dimensional field, *electromagnetic fields* exist in three-dimensions, and so *are three-dimensional fields*.

Since electromagnetic waves are made of photons, we conclude that the electromagnetic field (a wave) plays a (roughly) similar role for them as does the wavefunction for electrons. Thus, the electromagnetic field is a *quantum field*. The behavior of this quantized field is explained by a quantum field theory called *quantum electrodynamics* (QED). The birth pangs of the photon concept lasted about two decades from Planck's 1901 discovery, and it wasn't until the 1940s that the complete theory of quantum electrodynamics was successfully formulated.

Let us review the essential features of the electromagnetic field. The energy of the electromagnetic field is not spread out like a classical wave, but instead is concentrated in quantized *bundles* — the photons. For a given frequency, each photon has the same tiny amount of energy. Strong fields have very many photons, each with that tiny amount of energy, and weaker fields have correspondingly fewer photons. The photons follow probabilities that are determined by the electromagnetic fields.

Chapter 6

QUANTUM MECHANICAL REALITY

1. BATTLE OF THE GIANTS

> On this virtual field, the old-timers still tell
> that two Giants contended armed only with *thought*.
> And their *logic* as sword and as shield served them well,
> as with brilliance and fury, the battle was fought.

Einstein saw, soon after quantum mechanics was proposed, that it imbued nature with weird characteristics, besides being indeterministic. In a landmark paper by Einstein, Podolsky and Rosen, the ramifications of quantum mechanics in a simple experiment are carefully discussed. That *thought* experiment is commonly referred to as "the Einstein, Podolsky, Rosen" or "EPR" experiment (or "paradox"). These authors considered two particles (1 and 2) that interact for a short time, after which they move apart. If we know the quantum state of the particles while they are interacting, quantum mechanics will predict the quantum state after they have separated, no matter how far apart they are. If we then measure the *momentum* of particle 2, we can calculate the *momentum of particle 1* without disturbing it! Consequently, E, P, and R argue, "*there is an element of physical reality corresponding to this physical quantity.*"* Alternatively, we can measure the *position* of particle 2, and quantum mechanics will allow us to infer *the position of particle 1* from that measurement without disturbing particle 1. Consequently, the *position* of particle 1 must be an element of physical reality as well. So, here are two elements of physical reality, namely the momentum and the position of particle 1, which quantum mechanics says cannot be simultaneously determined. Their conclusion was that quantum mechanics is, therefore, not a complete theory. This conclusion is referred to as *the EPR paradox*. Niels Bohr, one of the great physicists of the first half of

*These words, along with the *italics*, are from the paper "Can quantum-mechanical description of physical reality be considered complete?" by A. Einstein, B. Podolsky and N. Rosen, *Physical Review* **47**, p. 777 (1935).

the twentieth century, responded to this onslaught by showing that their conclusion was unfounded, because of a hidden ambiguity in the reality criterion presented by E, P, and R. That is the claim that a measurement on particle 2 did not disturb the "system," which they considered to be particle 1. In quantum mechanics, *both* particles comprise the system, so that the measurements *do* disturb *the system* and the EPR paradox is resolved.

The instantaneous effect on particle 1, of a measurement performed on particle 2, is a non-local effect that is intrinsic to the quantum-mechanical behavior of matter; it is not caused by any interaction between the particles, but is a result of the *quantum-correlation* of their respective properties. This effect on particle 1 of a measurement of a property of particle 2 has been referred to as the "spooky action-at-a-distance" property of quantum mechanics. This *ghostly* non-locality was found objectionable by a number of physicists in the first half of the twentieth century, just as it disturbed E, P, and R. However, as *they* admitted in their paper, philosophical descriptions of reality should be based upon observations and experiments, rather than on predisposition or belief. The idiosyncratic properties of quantum mechanics, including *spooky and ghostly* effects, were experimentally verified in many different types of experiments during the last thirty years of the twentieth century and on into the twenty-first.

Albert Einstein and Niels Bohr had many discussions over many years concerning the self-consistency of quantum mechanics and its implications for epistemology ("the study of the origin, nature, methods and limits of human knowledge," from my trusty Webster's Unabridged). During meetings of the leading physicists of their day, they often stayed at the same boarding house (with several other attendees). At morning breakfasts, Einstein would pose another apparent contradiction resulting from quantum mechanics; by evening Bohr was able to refute the argument. In the end, Einstein was finally convinced that quantum mechanics was logically possible; however, it strongly violated his scientific instinct, which is expressed in his statement concerning "God" and "dice."

2. THE TRIUMPH OF QUANTUM MECHANICS

Bell's inequality

Can quantum mechanics be merely an approximation to a *deterministic* theory that has *no* mysterious action-at-a-distance ("non-local") effects? If so, there would have to be properties in that theory that are, so far, *hidden* from us and do not show up in quantum mechanics. Such a *local hidden-variable theory* would certainly bring equanimity and contentment to those who abhor a probabilistic, as well as a mysteriously non-local reality.

In the course of contemplating the EPR paradox and instantaneous reactions at a distance that are precluded in local hidden-variable theories, John S. Bell

found an unexpected result. He showed that viable* local hidden-variable theories would *all* contain a particular relationship among certain measured quantities in particular types of experiments. This relationship is now referred to as *Bell's inequality*. Its form was: *The sum of a certain set of results must be less than the sum of another set* in *any* local hidden-variable theory. The remarkable feature of this inequality was that, in some experiments, *the quantum mechanical predictions violate the inequality*. So here is a quantitative *experimental* way to distinguish between deterministic, non-spooky (local) hidden-variable theories and probabilistic, weirdly non-local quantum mechanics; measure the quantities and form the sums. If the results violate the inequality by producing the quantum mechanical prediction, then all local hidden-variable theories are invalidated at once! Subsequently, several groups of physicists derived alternate versions of Bell's inequality suitable for several *different* types of experiments. *All* the experimental results were in agreement with the predictions of quantum mechanics and consequently violated the inequalities required by *any* local hidden-variable theory. *The world is spooky after all!*

<div align="center">

Reality

(to the "Ode to Joy" theme in the Fourth Movement of Beethoven's Ninth Symphony, with apologies to lovers of F. Schiller, L. v. Beethoven, A.Einstein, N.Bohr, and/or J.S. Bell)

Al and Niels had long discussions
'bout the myst'ry — reality.
They did not foresee the coming
of Bell's inequality.
Al and Niels had such long discussions
about the quantum the-ory.
They explored, illuminated
the nature of reality.

</div>

"Schrödinger's cat" exists

Erwin Schrödinger, an Austrian physicist, was one of the major players in the creation of quantum mechanics. In 1926 he created the wavefunction and the wave equation (which bears his name) that it obeys. (That equation and its wavefunctions are at the heart of almost all quantum mechanical calculations in solid-state and atomic physics.) However, he was not comfortable with the *probabilistic* nature of quantum mechanics, and in 1935 suggested a rather ludicrous thought experiment to shock as well as cajole its proponents. He was trying to show how *ridiculous* such a state would be for macroscopic objects. In this "experiment," a cat is placed inside a box containing a vial of poison and a radioactive substance. If the substance decays, the particle it emits would set off a chain of events that would break the vial releasing the poison and killing the cat. Suppose we wait one half-life (of the radioactive substance) before opening the box, that "opening" being the presumed

*whose results agree with those quantum mechanical predictions that had already been experimentally verified.

measurement. The probability that the cat will be dead is $\frac{1}{2}$. Should we believe that the cat is neither dead nor alive before the measurement, i.e. before the box is opened? This *metaphysical* question was ignored in the *practice* of the science of physics until the 1980s, although it had stimulated conjectures about the nature of reality. From the 1980s on into the 2000s, several experiments were done that produced (microscopic) states of atoms or molecules in which the wavefunction had a part corresponding to the particle being at one place and a part corresponding to it being at another. In other words, the state had a *probability distribution* for the location of the particle. Of course once a measurement of its location is made, the particle would be found at one place *or* the other.

The state whose wavefunction contains a distribution of possibilities is called a "coherent superposition" (of those different possibilities). We have, in fact, already discussed such states and they have been created in many experiments even early in the twentieth century, although this coherent-superposition property was not emphasized (or reported to the NY Times). They are the states of a particle (like an electron or proton) with spin up (or one with spin down) in a given direction. Consider a particle that is spin up in a particular direction. Quantum mechanics shows that a measurement of its spin in a direction *perpendicular* to that has equal probability of being spin up or spin down; it is thus a superposition of the state spin-up in that perpendicular direction and the state spin down in that direction. No big deal!! No news conference. Furthermore, looking at Heisenberg's Uncertainty Principle, we see that every time we produce a state with a very narrow range of one observable (like momentum), that state is a coherent superposition of states with a wide range of definite values of the conjugate observable (like position, and vice versa).

Now let us go back to the states that are coherent superpositions of the *location* of an object. Should we describe such a state by saying that the particle was at two places at the same time, or was it no place? Of course, for NY Times, etc., stories it was claimed that the particle was at two places at the same time! WOW! However, we might say instead that being *forced* to describe a coherent state as being at two places simultaneously or no place at all is inappropriate. Classical language is not always appropriate for quantum phenomena.

It has been pointed out that the *only* reason that we do not see macroscopic objects (like you, me, or "a cat") in such coherent states is because any interaction of an object with its surroundings is equivalent to a *measurement* that throws it into a state of definite location, etc. Such an occurrence is called *decoherence*. The implication for Schrödinger's cat is that the radioactive substance will either decay or will not (because of its interactions with its surroundings), but will not remain in the state that contains both probabilities for any practical length of time. Therefore the cat will be either dead or alive, but will not be in the unimaginable state containing both probabilities, as we open the box. However, even early in the twentieth century, people had succeeded in creating *macroscopic* systems at

very low temperatures that each remained in a spectacular quantum state without interacting with their environs. These were the famed superconductors.

More recently, states have been observed in which an atom is here and there at the same time, in the sense of a coherent state containing both possibilities — but *microscopic*. And, in the year 2000, a group of scientists at The State University of New York, Stony Brook succeeded in putting a superconducting loop into a coherent superposition of two *macroscopic* states — one with the current flowing clockwise around the loop, the other with the flow counterclockwise. The loop was in the shape of a square, with sides about 5 micrometers in length (about the width of a human hair), which is indeed very small but still macroscopic — the size of tens of thousands of atoms laid end-to-end. So "Schrödinger's cat's" coherent superposition really exists and the eerie quantum-spookiness of nature has been demonstrated in an object big enough to be seen and touched!

Delayed-choice experiments

In recent years, experiments have been performed which show either the wave or the particle nature of light, depending on a last-minute choice of the final detector. These experiments directly verify the instantaneous non-local behavior of nature as predicted by the inherent structure of quantum mechanics. In these *delayed-choice experiments*, the final detector is inserted *after* the light has gone through the rest of the apparatus. Let us consider their basic structure: Instead of two slits, a half-silvered mirror is used to break a laser beam into two beams, one reflected and one going right through the glass, as shown in Fig. 6.1.

The reflected beam is reflected again by another mirror, so that it crosses the transmitted beam. In the crossing region of the two beams, we can put a detector (1) that shows *which direction the detected photon came from*, (*e.g.*, two beam pipes with an electric eye in each), or a different detector (2) that shows *interference effects of the two beams* instead (*e.g.*, a screen). This interference can occur only because of the existence of *both* beams, so that the electromagnetic wave producing an interference pattern on detector 2 must be the result of "waves" that went *both ways*. When the choice of detector is made, the *light to be received has already left the region of the splitting of the beam*. So how can this *delayed* choice of detector affect the behavior of the light at the *earlier* splitting (a photon going only one way, if detector 1 is used; or "waves" going both ways, if detector 2 is used)? — When detector 1 is used, it *does* show that the light detected in the beam-crossing region was from one beam *or* the other, but when detector 2 is used, it shows instead that the detected light was *partly from each* beam. The answer that quantum mechanics has for this apparent paradox is that *the entire set-up including the detector* must be considered before we can conclude anything about the experimental results. The light will appear to have gone only one way when we use detector 1 and both ways when we use detector 2. We must think of the

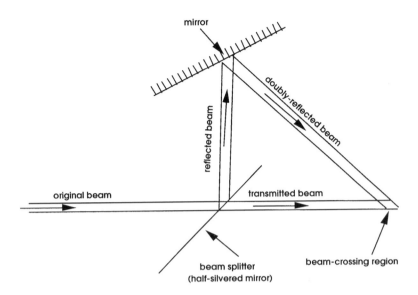

Fig. 6.1. Basic delayed-choice experiment. The original beam is shown entering from the left. A detector is placed in the beam-crossing region, just before the transmitted and doubly reflected beams arrive.

detector as part of the full quantum system. Before a detector is inserted, the quantum state has both outcomes as possibilities and the insertion of the detector changes the *entire* quantum system to one or the other outcome. In classical physics, which is more in line with our intuition, each part of a system has a separate *closed* reality — first one thing happens then the next, etc. However, quantum mechanics tells us that we cannot chop up a system and consider it piece & instant by piece & instant. There is, instead, *a fundamental* non-locality (in both space and time) in nature, that requires us to consider the whole system and the entire interaction involved in a particular chunk of experience (*e.g.*, in an experiment). There is even a more extreme interpretation that proclaims that nothing exists until it is measured, but this would take us too far from our paths to explore here.

Nature's verdict

So which wins, our intuition or quantum mechanics? Remember that our intuition was presumably formed during our infancy from our interactions with ordinary macroscopic (large compared to atomic sizes) objects. In our discussion of relativity, we have already seen that our intuition is wrong about objects moving with speeds near the speed of light; now it appears to be mistaken about the nature of reality as well. Here we are talking about the reality we understand from our interactions with nature, often referred to as *operational reality*. We have nothing to say about the ultimate, perhaps unknowable, REALITY. (Recall the discussion in the first chapter.)

Trumpets: TAH-TAAH
ALL HAIL THE TRIUMPH OF QUANTUM MECHANICS

3. QUANTUM MECHANICS AND MEASUREMENT

Surprisingly, not all physicists agree on what actually happens *during* a measurement. The *words* used to describe the action of measurements on the wavefunction vary according to which theory of measurement the scientist speaking embraces. In the traditional "Copenhagen Interpretation," a measurement actually "collapses" the wavefunction to another mathematical form whose properties conform to the result of the measurement. The original form of that interpretation considered "the measurement" to include the conscious mind that becomes *aware* of the result, because the wavefunction was considered a representation of our knowledge of the system. (This leads to the dilemma of whose consciousness determines the wavefunction, the person's or the cat's,* in the "Schrödinger's Cat" scenario.) We will not investigate that aspect of the interpretation, since it is far off of our itinerary. Instead, we observe that it is *experimentally* verifiable that the wavefunction has been changed by the measurement — which can be operationally called a "collapse," thus bypassing the creation of a scenario for the "during" period. Despite the *differences* of the descriptions of the measurement process as it changes a state, the equations of the theory and their predictions are embraced by *all* scientists, and the probabilistic nature of *nature* is widely accepted.†

*especially if the cat is replaced by a second person.

†Theories *have* been proposed that turn quantum mechanics into a *deterministic* theory, but require the existence of an infinite number of worlds. These Many-Worlds theories reside in the realm of *metaphysics* and a discussion of them would take us very far from our itinerary!

Collapsing at a Restaurant*

(with apologies to vegetarians, Lord Randal and his mother)

The Wavefunction:

Lord Randal: *Let's see, the salmon does sound good,*
but I might get shrimp scampi.
Oh, look they have stuffed cabbage here,
stuffed grape leaves, lamb chops, swordfish

The Measurement:

Waitress: *So what is your order, my handsome young man?*

The *Collapse* of the Wavefunction:

Lord Randal: *U..um, OK, I'll have shrimp scampi!*

The Next Measurement:

Lord Randal's mother: *And what are you having Lord Randal, my son?*

The Collapsed Wavefunction:

Lord Randal: *Shrimp scampi, yes, shrimp scampi!*

 (The waitress leaves, and later returns with the food.)

The Next Measurement:

Waitress: *Now which is your order, my handsome young man?*

Again The Collapsed Wavefunction:

Lord Randal: *Shrimp scampi, yes, shrimp scampi!*
.
.
.

Lord Randal: *I'm weary wi' eating and fain wald lie down.*

(collapse)
(See the last stanza below.)

*inspired by the old English ballad *LORD RANDAL:*

"O where hae ye been, Lord Randal, my son?
O where hae ye been, my handsome young man?"
"I hae been to the wild wood; mother make my bed soon,
For I'm weary wi' hunting and fain wald lie down."

"Where gat ye your dinner, Lord Randal, my son?
Where gat ye your dinner, my handsome young man?"
"I dined wi' my true love; mother make my bed soon,
For I'm weary wi' hunting and fain wald lie down."
.

(his bloodhounds "swell'd and they died")
.

"O I fear ye are poison'd, Lord Randal, my son!
O I fear ye are poison'd, my handsome young man!"
"O yes! I am poison'd; mother make my bed soon,
For I'm sick at the heart, and I fain wald lie down."

Now we take our first look at the elementary particles and their interactions. There are a very large number of particles in the category we call "elementary;" however, they are composed of only a handful of fundamental particles. Those fundamental particles and their four basic interactions cause everything physical we have ever encountered — all of the interactions and diverse materials observed in nature result, we believe, from the interplay of these underlying entities.

Chapter 7

THE PARTICLES OF MATTER

1. SEEKING THE STRUCTURE OF PROTONS AND NEUTRONS

In the past century, we have explored the structure of smaller and smaller particles in our intriguing quest for the basic building blocks of matter. Since the ground breaking scattering experiments of Rutherford in the early part of that century, it has become usual to fire rapidly moving particles at target particles in order to search for structure in either or both particles. By colliding a particle whose properties we understand with one that is under investigation (or two identical particles) and analyzing the resulting particle "debris," we should be able to "see" what was originally inside. (Shining a light to look inside something is analogous to firing *photons* at a particle under investigation, since light is, in fact, made up of photons.) HOWEVER: When the proton was investigated by means of scattering experiments, it did not break into "constituents," instead *completely new, unexpected particles were created!*

Actually, particle creation had first been observed earlier in the twentieth century (1930s). In many experiments, a high-energy photon would disappear with the concurrent appearance of a pair of charged particles (an electron and a positron) in its place. This "creation" of a particle pair from a photon was understood in terms of *quantum electrodynamics*. We certainly do not think that the photon is *composed* of an electron and positron, but instead the electron and positron are fundamental massive matter particles and the photon is a massless particle of light. Quantum electrodynamics explains this photon annihilation and pair creation in a natural way, right from the mathematics of the theory; however, in the proton-proton scattering experiments the particles created are *new* ones, previously unknown.

Scattering?

(Inspired by the song "Comin' Though the Rye" that starts "When a body meets a body....")

When a proton meets a proton, flying fast and free.
Their collision is stupendous; on that we all agree.
As they smack into each other, there are particles begat.
Now let me emphasize again, they do not just go "splat!"

Imagine that!

A very large number of new particles have been created in the scatterings of the types of particles we find around us (*e.g.*, protons, electrons and photons) and beams of some of these newly discovered particles have been used to produce more new types of particles. In all cases it can be concluded that the newly produced particles were not *inside* the colliding particles. In fact, hundreds of such new particle types have been created in this way... *and then there were many? very many!* Could *all* of these be fundamental particles? Might some of them be composites of the others, or are some or all, perhaps, composed of more fundamental constituents?

In these scattering reactions, we usually see more total mass coming out than went in. How could more mass emerge than was present before the scattering took place? Well, there is no conservation of mass law! Einstein's theory of relativity has taught us that mass is a form of energy (rest energy) and so, as long as the initial particles are moving fast enough to supply the extra mass-energy, these reactions *do* occur. Some of the motional (kinetic) energy of the initial particles is converted into mass by the reaction, conserving the total energy. (It is just the opposite of what happens when the uranium in a reactor fissions, where that massive nucleus produces nuclei with less total mass but moving at high energies.) Millions of such reactions have been seen and analyzed in rays of particles that are continually bombarding the earth from all directions (cosmic rays), and at huge particle accelerators that have been built to perform controlled scattering experiments.

2. STRONGLY INTERACTING MATTER PARTICLES — THE HADRONS

The particles in the nucleus of an atom, i.e., the protons and neutrons, are relatively heavy particles which have a strong interaction with one another, sometimes referred to as the nuclear force. In collisions of high-energy cosmic rays and in numerous experiments, we find "cousins" of the neutron and proton, which are even heavier strongly interacting particles. All of these strongly interacting particles are called *hadrons*. We now know that they are all made of the fundamental particles called *quarks* that interact even more strongly; however, those constituents are "locked" inside the hadrons. We shall have a lot to say about the constituents of the hadrons, their interactions, and how their presence was discerned later.

There are two different kinds of hadrons, whose quark structures are quite different: The proton and neutron and their heavier strongly interacting cousins form

one type and they are called *baryons* (from the Greek for "heavy"). Those heavier than the proton or neutron decay rapidly, leaving a proton or neutron to attest to their fleeting existence. Even the neutron decays, though it lives about fifteen minutes* before changing into a proton plus lighter particles. So, all of the heavier baryons eventually disappear leaving a proton, the lightest baryon, in their place. None of the baryons is a fundamental particle, but instead, they are all composed of 3 quarks. Their antiparticles are called *antibaryons* and are made of 3 antiquarks.

Particles of the other type of hadron are called *mesons* (from the Greek for "middle"), since the masses of the first ones discovered are midway between those of an electron and a proton. They *all* decay rapidly, ultimately to photons or to electrons, positrons and other light particles that do not interact strongly. Mesons are each composed of a quark and an antiquark.

There are many *hundreds* of different types of hadrons produced in high-energy scatterings. Although they still retain the designation *elementary particle*, we now know that not one of this plethora of hadrons is a fundamental particle! However, the quarks inside them are so tightly held there by (strong) cousins of the photon called "gluons" that they can't get out. Nevertheless, we shall soon see that (conceptual) symmetry groupings of these hadrons according to their various properties provided the first hints of the existence of quarks.

Hundreds of Hadrons

Hundreds of hadrons were found in the hunt.
Hundreds of hadrons, we had to confront.

Two happily hurrying particles, free,
helplessly hurtling towards catastrophe.
How could they know what their fate had to be,
hitting each other at *high* energy?

Hastily vanishing heroes, those two,
handed *new* hadrons a hearty debut,
hundreds of hadron-types, for us to view
hints of *constituents* — what a huge clue.

Honor and glory are long overdue!

Hip - Hip - Hooray!

3. MATTER PARTICLES WITH NO STRONG INTERACTIONS — THE LEPTONS

The electrons inside the atom are not confined to the tiny region inside the nucleus; they do not feel the nuclear force at all. We express this by saying that they do not interact *strongly*. However, when a radioactive atom decays, an electron or

*That is roughly its *mean* life. Its half-life is 10 minutes.

positron emerges from the nucleus. In fact a neutron, when it is not inside a nucleus, also produces an electron as it changes into a proton. The electrons or positrons emerging from radioactive nuclei or from decaying neutrons are being *created just as they emerge* by an interaction much weaker than the strong interaction, called *the weak interaction*.

The emergent electrons and positrons in these reactions were first referred to as beta (β) rays because no one knew their composition. In the 1930s, analysis of these so-called *β-decays* found that the conservation laws of energy, momentum and angular momentum *appeared* to be violated by the data. Some very prominent physicists of the time conjectured that the weak interactions may be giving us our first glimpse of the violations of energy and momentum conservation. HOWEVER, a strong belief (uh-oh) in these conservation laws, which are *required* by our basic theoretical understanding of the physical world, produced the conjecture that an unseen, uncharged particle had emerged from the nucleus along with the electron. *Careful analyses of very precise experiments* showed that this was indeed a viable assumption, and the ghostly particle was named the *neutrino* (Italian for "little neutral one"). These uncharged particles could not interact strongly or electromagnetically, only weakly; this would account for their undetectability in those β-decay experiments. In the 1950's, gigantic neutrino detectors were built which did indeed *find* these particles. So, hunches and intuitive guesses sometimes reap great rewards! We now know that the electron and the neutrino associated with it (and their antiparticles) are *weakly interacting fundamental particles*.

Subsequent experiments over a period of several decades have shown that there are two more, heavier "copies" of the electron. The first one found is called the *muon* (pronounced: m-you-on) or mu for short and is represented by the Greek letter μ (mu). The other is the *tau* (pronounced like the first syllable of the word "tower") and it is represented by the Greek letter τ (tau). They each have their own corresponding neutrinos. The discoveries of these heavier versions of the electron came as complete surprises — there were no hints from theory of their existence.

Why the Tau Has No "on"

You'd think charged leptons of repute would have a little "on,"
for that would light their way into society's bastion.
But let us not go on and on 'bout "on" and its fair place.
We'll get right to the matter of this *matter's* great disgrace.

"The tau avoided contact with our species, out of spite!"
So *they* said they would punish it, and thought they were quite right.
But it was not the tau at fault, for being hard to find.
Its giant mass leads to decays of many different kinds.

The ven'rable electron has its "on" on as its *creed*.
The muon sometimes loses it, but not when there's a need.
But the poor tau is ostracized, an "on" it cannot use,
and so polite society's acceptance is refused.

Now let us sing the song of tau, with wailing of great woe.
The heaviest of all leptons should not be treated so.
Its dignity, in spite of all of this, stands proud aglow,
as we now sing this song of tau, with wailing of great woe.

Carry *on*, tau!

(This poem was inspired by two reviewers of the technical book I wrote on this subject, who objected to my use of "tauon" in context with "electron" and "muon." However, I recall a colloquium in which the discoverer of the tau used the word "tauon," and it has appeared in popular books as well.)

The muon is indeed a *heavy* copy of the electron; it is about 200 times as massive as the electron, and decays rapidly into an electron and neutrinos. The tau is enormous being 17 times as heavy as *the muon*, which amounts to about two times the mass of a proton. With all that mass, the tau has many different decay modes. The three corresponding neutrinos are presumed to have exactly zero mass in our Standard-Model theory of elementary particles. However, several experiments indirectly show that their masses probably differ from zero, but by extremely small amounts. We expect to have a definitive answer to the question of neutrino mass in the first decade of this new century.

All together, these weakly interacting particles are referred to as *leptons* (from the Greek for "thin", because those types *originally* found have very low mass). The tau, at twice the mass of the proton, doesn't seem to qualify for such a label, but its interactions mimic those of the electron so it is, indeed, classified as a *lepton*. Just like the electron, the other leptons each have spin $\frac{1}{2}$.

The leptons are often grouped into pairs: electron (e^-), electron-neutrino $(\nu_e)^*$; muon or mu (μ^-), mu-neutrino (ν_μ); tau (τ^-), tau-neutrino (ν_τ); and their antiparticles. (The minus signs indicate their charges.) Each of these three pairs is considered a separate lepton *generation*. The first generation (the electron and its

*The letter 'ν' is the Greek letter pronounced 'noo;' we call this 'ν sub e' or 'ν e.'

neutrino) occurs in the matter all around us. The second generation particles (the muon and its neutrino) appear as secondary products when cosmic rays smash into molecules of the atmosphere. We have found the third generation of leptons (the tau and its neutrino) only in high-energy accelerator experiments, so far. The tau decays so rapidly that we are only able to detect one right near its creation site.

In our travels, we shall learn that the fundamental *strongly* interacting particles, the quarks, also form three generations with two individuals in each.

Quark Bondage

(inspired by the rhythm of the musical introduction to the old children's program "Mr. Ed")

A hadron is made of quarks, of course:
A meson has quark and antiquark;
Three quarks bound together 's a baryon.
The quarks, in these bound states, sure carry on!

The gluons, they grip the quarks so tight
that never a one comes into sight.
And just as the quarks, for life, are bound,
their jailers, the gluons, 're never found.

Lepton Freedom

(to the music from "Porgy and Bess" with apologies to the Gershwins and D. Heyward)

Oh, I got plenty of leptons.
These leptons' plenty for me.

But ... got no tau, got no mu
(got no neutrinos).
Those ... two generations of leptons,
ain't here, ain't there, 'r anywhe- ere.
Seems with plenty,
you sure got to worry
that they'll soon all decay away ... away.

I got no lock on the door,
that's no way to be.
The first generation of leptons
ain't gonna decay
And the things, that I prize
like neutrinos from the skies
are all free....

Oh, I got plenty of leptons.
Electrons are plenty for me.
I got my hadrons,
got my leptons,
got atoms all along.
(No use complainin'!)
Got my hadrons, got my leptons,
got my song.

Oh, I got plenty of leptons.
These leptons 's plenty for me.
They're not like the quarks
that are bound in my hadrons;
instead, they're essentially free,
like me!

Chapter 8

THE BASIC INTERACTIONS

1. INTRODUCING THE INTERACTIONS

Although we have encountered some of the basic interactions during our visit with the fundamental particles, we have not been formally introduced. Let us begin our introductions by taking a closer look at the ubiquitous atom. When we explore its tiny central nucleus, the hard-candy center of our colossal cotton-candy treat, we find that it is made up of even smaller constituents just as a time-release capsule has those tiny spheres inside. There are two kinds of particles inside the nucleus — protons and neutrons, which are held together by the *nuclear force*. This force does not depend on electric charge at all; it is exerted between protons and protons, neutrons and protons, as well as neutrons and neutrons. It is so strong as to overshadow the electrical repulsion of the protons (by one another) in the nucleus, which would blow the nucleus apart were there no nuclear force; however, the nuclear force has a very short range. On the other hand, the "cotton-candy" electrons do not feel nuclear forces at all. They are held in the atom by the much weaker, but long-range, *electrical forces*; the negatively charged electrons are attracted by the positively charged nucleus (containing protons and neutrons). The nuclear and electromagnetic forces are manifestations of two *basic interactions* of nature, *the strong and electromagnetic interactions*, respectively.

In our everyday lives, most of our experiences result from the electromagnetic interaction, for all chemical interactions, including the biochemical interactions going on in our bodies, are electromagnetic interactions among the electrons of different atoms. On the other hand, we become aware of the nuclear interactions only when we read about nuclear power plants (or regrettably weapons), are reminded that the sun's and other stars' sources of energy are nuclear reactions, or have some unfortunate acquaintance with nuclear medicine.

We are also informed by the media that some atoms are radioactive, with particles (radiation) emerging from the nucleus. Many of these radioactive decays of nuclei are caused by a third, weaker type of basic interaction called the *weak*

interaction. This interaction has been technologically exploited in medical and industrial uses of radioactive substances. It is also a source of our concerns about nuclear waste.

The fourth basic interaction of nature is the *gravitational interaction*, which is manifested in our everyday lives by the pull of the Earth, that we usually refer to as *gravity.* The gravitational interaction has played a major role in the evolution of the universe, from its presumed simple but cataclysmic Big-Bang origin to its present day complexity.

And that is it! There are a total of *only four* basic interactions in nature, in terms of which all observations of natural processes presumably can be explained. Here we purposely use the word "interaction" rather than "force" because the interactions among particles often involve the *annihilation* and/or *creation* of particles — not just pushes and pulls.

<div align="center">

Those Mysterious "Interactions"

"Interaction" is not just a force.
For particles *can* disappear,
just as others burst forth from a source
that's beyond us, beyond our thin tier.

A force can push, a force can pull
but can't explain that mystery.
It can't enshroud a particle.
Nor can it conjure one to *be.*

</div>

Let us pause now, to elaborate on what we have seen here. The basic interactions are, in order of decreasing strength:

- the *strong interaction,* which is responsible for the nuclear forces holding the nucleus together. It is the strongest of the basic interactions but acts only over a *tiny* distance, which is roughly the size of nuclei.
- the *electromagnetic interaction,* which is responsible for the interaction of the electrons in an atom with the nucleus (and among themselves); it holds the atom together. It is also responsible for the interactions *between* atoms known as chemical reactions. All electrical, magnetic, and optical phenomena result from this interaction. The emission of photons* from the nucleus involves this interaction as well. It is about one hundredth as strong as the strong interaction, but has an exceedingly long range.
- the *weak interaction,* which manifests itself in radioactivity, namely, the sudden appearance of electrons and neutrinos emanating from the nucleus, known as beta (β) decay. The weak interaction has been measured to be only about one one-hundred-thousandth as strong as the strong interaction, and to have an *extremely* short range, roughly one thousandth the typical size of a nucleus.

*sometimes referred to as γ (gamma) rays.

- the *gravitational interaction*, which is responsible for the "gravity" forces between massive objects. This interaction has an exceedingly long range, but it is very, very,..., very weak. The extreme weakness of this interaction seems incredible, but is *not* inconsistent with the large gravitational forces we see around us, and the major role that gravity plays on the cosmological scale; these large gravitational forces are a result of the fact that gravity is *solely an attractive force* so that for very large objects there is a large total force. This contrasts sharply with the other long-range force, the electric force, where like charges repel and opposite charges attract, so that many, many cancellations occur for large objects. In the vast majority of the interactions considered in particle physics, where we are considering a very small number of particles (as in the atom), the gravitational interaction is so small as to be unobservable.

2. THE GAUGE* BOSONS OF THE INTERACTIONS

It is now time to meet the *gauge bosons* (*or gaugons*). The strong, electromagnetic, weak, and gravitational interactions are each believed to be transmitted between particles by a "virtual" form of these gaugons. These gauge bosons, when playing their roles as *virtual particles*, are not present as physical particles; instead, they are like *phantom arms, with hands at both ends* that reach from one particle to the other. Sometimes a particle's type is actually changed by its interaction with a virtual gauge particle. The action of virtual gauge particles is represented pictorially in Fig. 8.1.

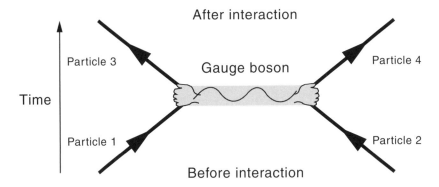

Fig. 8.1. Symbolic representation of the interaction of particles (1) and (2) via a virtual gauge boson, a "phantom arm" whose "hands" are "grasping" each particle. The emerging particles, (3) and (4), may be different types from the incident ones.

*rhymes with "page."

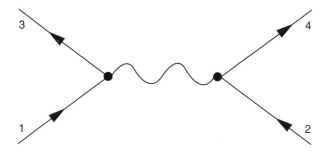

Fig. 8.2. Feynman diagram for the same interaction represented pictorially in Fig. 8.1.

With enough energy, *real* versions of the gaugons can be produced so that they can be directly studied.

The drawing in Fig. 8.1 is an elaboration of the kinds of diagrams invented by Richard Feynman (the first syllable is pronounced like the word "fine"), with which he masterfully developed and explained quantum electrodynamics. In his diagrams, the virtual particles are not arms and they do not have hands, but instead the diagrams contain only the essentials. Figure 8.2 shows the corresponding Feynman diagram.

Each of the two points in the diagram that represent the joining of the gauge particle to a matter particle is referred to as a *vertex*. Feynman's approach was to consider the virtual particle as having been emitted by one of the particles and absorbed by the other, and then including both possibilities. Consequently, the virtual gaugon is often referred to as the "exchanged" particle. Actually, the calculation corresponding to the diagram in Fig. 8.2 describes the process only to a certain accuracy. For more and more accuracy we must also calculate the diagrams with larger and larger numbers of virtual particles. The results so obtained are called "perturbative* solutions." (Non-perturbative methods that do not look at individual Feynman diagrams have been developed for special situations).

And what is the rationale for the name "gauge boson?" The word "gauge" appears because the existence of the exchanged particles is explained by the *gauge theory* of elementary particles called the "Standard Model," which will be explored in Part III of our journey. The word "boson" appears because they have integral spins (actually, spin 1 or spin 2).

There is a different *kind* of gauge boson for each interaction. Let us greet them, now:

*usually pronounced with the accent on the second syllable and the 'a' like that in 'attire.' The word was originally used for a series of corrections to a *known* solution due to an interaction that *perturbs* the original system — e.g., calculations of planetary motions, whose paths were perturbed by the pulls of the other planets from that due to the sun alone, and quantum mechanical calculations for an interaction that perturbs the original system.

- Inside the hadrons dwell the *gluons*, the gauge particles of the strong interaction, which are massless particles. These provide the extremely strong "gluing together" of the quarks in the strongly interacting particles (hadrons). The gluons have never been found *directly*, nor do we expect to be able to produce or detect free gluons. However, we shall see later that direct *evidence* of their existence *has been found.*

<u>Gluons</u>
(Inspired by the folk song that starts "Freight train, freight train...")

Gluons! Gluons, holding so fast.
Gluons! Gluons, there is no blast
that can free the quarks you hold,
as all color you enfold.

Gluons! Gluons, holding so tight.
Gluons! Gluons, it is your plight
to be stuck inside as well,
in the same strong color spell.

- Now greet our old friends the *photons*; they are not the least bit shy. The photons are the gauge particles of the electromagnetic interaction. They are also massless particles. These particles comprise light and all the other electromagnetic waves; they have been detected as separate particles in very many experiments.
- The gauge particles of the weak interaction are the W^+, W^- and Z^0, which have very large masses, roughly 80 (for the W's) and 90 (for the Z) times the proton mass. In their virtual guises, they mediate the weak interactions. These particles were found as *real* particles in the early 1980s by a collaboration of hundreds of physicists, in a huge particle accelerator in Geneva, Switzerland.* Such a colossal experimental effort was required because of their large masses.

W 'n Z
"*W* 'n *Z*, *W* 'n *Z*,
what can you two possibly be?
Why are you so very heavy?
C'mpared to protons, a mass near ninety!"

"Well now, here is the answer to your inquiry:
There are many more gaugons that you never will see;[†]
and compared to them we're not at all heavy!

Our int'ractions, you surely, without doubt agree,
are detected and studied very easily,
for the *weak interaction* is our repartee.

Oops! My time to decay has come flurriedly,
so I'll say my goodbye to you hurriedly."

*It is an internationally owned and operated facility called CERN (Conseil Européen pour la Recherche Nucléaire).

[†]Theories proposed beyond the Standard Model predict the existence of very heavy gaugons, which are not detectable in present-day experiments.

- Gravitation is also believed to occur because of the exchange of particles: They are the massless *gravitons*, presumed gauge particles of the gravitational interaction. However, they have yet to make an appearance for anyone, so do not feel insulted by their absence. No direct evidence of gravitons has been found as yet, due to the inherent weakness of the gravitational interaction. Gravitational waves themselves have not been detected yet, let alone such evidence of their quantization. However, recent observations of a double neutron star system show evidence that gravitational waves are carrying energy away. So these waves probably do exist.

<u>Do You Exist</u>

(suggested by the rhyme "Oh, where have you been, Billy boy, Billy boy?")

Oh, do you exist graviton, graviton?
Oh, do you exist as expected?

We will seek you, without doubt
when our techniques can reach out.
Though a shy thing, you'll surely be detected!

3. THE "REACH" OR RANGE OF THE INTERACTIONS

Now, let us take a closer look at the distances over which the interactions operate. The effective distance of an interaction is called its *range*. Virtual particles can *reach* or *travel further* (whichever metaphor you prefer) before being reabsorbed, *the less massive they are*. So, the range of the interaction is usually (inversely) related to the mass of the corresponding gauge boson.

The *masslessness* of the photon and the putative graviton account for the *exceedingly long ranges* of the electromagnetic and gravitational interactions. Likewise, the *very short range* of the weak interaction can be explained by the *extremely large masses* of the Ws and the Z^0 gauge bosons, which "propagate" this interaction.

The strong interaction is an exception to the rule that connects the range of the interaction inversely to the mass of the corresponding gauge particle. The gluons, the gauge bosons of the strong interaction, have no mass, yet the *observed* range of the strong interaction (e.g. nuclear forces) is not long, but instead is *very short*. There is a much more subtle mechanism operating here, which we shall explore later.

Part II

PARTICLE SYMMETRIES

<u>A Song of Symmetry</u>

Let's sing a song of symmetry
with voices sweet and clear,
its poignant blending harmony
enchanting to the ear.

Let's sing a song of symmetry;
Legato lines we'll sing,
with covered tones of great beauty
whose fortes soar and ring.

Let's sing a song of symmetry
that echoes in our minds,
inspiring thrilling ecstasy
about the world's design.

Oh symmetry, oh symmetry
your beauty does astound,
and all of us are filled with glee
whenever you are found.

Now we begin our exploration of the concept of symmetry. It is here that we will find the crowning achievement of elementary particle physics in the twentieth century, namely the Standard Model. We will have to pass through several "domains" to reach it, for we must first acquaint ourselves with the concept of symmetry itself. We will pay special attention to the generators of continuous symmetries, as we are introduced to their associated conserved quantities.

Chapter 9

SPACETIME SYMMETRIES

Let's imagine going to an art fair! The weather is great, sunny but not hot. There are occasional gusts of wind, but they are refreshing rather than annoying.

1. SYMMETRY AT AN ART EXHIBIT

Rotational symmetry

First we visit the modern art exhibit at the fair. Our attention is caught by a painting of a solid square, done in an intriguing color. It is framed in a very strange manner, inside a square black frame with a long pin through the center of the painting. It is "hung" on the wall by that pin. When we first look at it, it is perfectly straight on the wall; the top and bottom are parallel to the line where the wall meets the ceiling. But the gusts of wind have their effect and when we look again, it is rotated slightly about the mounting pin. With your keen eye, you easily notice that the picture is crooked and you ask the artist to straighten it. To your initial surprise, he begins to rotate it further, but you immediately realize that when it reaches one quarter of a complete turn it will be straight again, as shown in Fig. 9.1. We have inserted numbers in the corners to allow us to track their positions. These numbers are not part of the shapes, and when they are ignored, the orientation does appear to be unchanged.

Further quarter turns would have the same result, so this picture is symmetric under rotations of one quarter (90°), one half (180°), three quarters (270°), and of course a complete rotation (360°, which is equivalent to no rotation at all (0°) since the picture is back in its original position). We see that this *group* of rotations is composed of a *separated, finite* number (4) of operations. Because of the separated nature of these operations, it is called a *discrete* symmetry group. A figure with a larger number of sides would be symmetric under a larger number of rotations and so its discrete symmetry group would be larger — the corresponding symmetry group of a pentagon (with five sides) would contain five rotations.

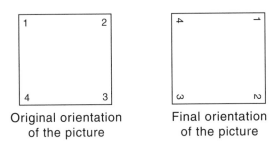

Original orientation Final orientation
of the picture of the picture

Fig. 9.1. Rotation of the square picture, clockwise a quarter of a turn (90°). The corners have been numbered here, so that we can keep track of their successive locations. (There are no numbers on the picture we are talking about.) Ignore the numbers, and you can see that *the two orientations look identical.*

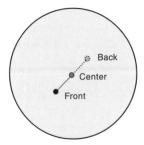

Fig. 9.2. A sphere with one of its diameters drawn in. A diameter is a line running from a point on the surface of the sphere, through the center, to the opposite point on the surface.

Now look at the painting over there, by the same artist, that is circular with circular brush strokes and a circular frame. It could never look crooked! It is symmetric under *any* rotation about its central pin. Its symmetry group is very different from those we have just talked about — we could *continuously* rotate that picture and, whenever we stopped, it would look the same. So, its symmetry group contains an infinite number of rotations, which can be chosen from a set running continuously from no rotation (0°) to a full rotation (360°). Groups with a continuous set of operations are called *continuous* groups. This one is called the *two-dimensional rotation group*, because it includes *all* of the rotations (0° to 360°) of the two-dimensional plane of the picture. Note that all of those rotations are about the same axis, namely the pin perpendicular to the painting. The pin (or a line through it) is referred to as the *rotation axis*.

Look at that beautiful spherical sculpture in the center of the exhibit, as shown in Fig. 9.2.

When it is rotated, around *any* of its diameters by *any* amount, it looks unchanged. Its symmetry group is *the three-dimensional rotation group*. If we choose a particular diameter, then the rotations (all of which are in the three-dimensional rotation group) about that diameter form a two-dimensional rotation group. This

shows that the two-dimensional rotation group is a *subgroup* of the three-dimensional rotation group (in fact, there is a continuum of choices of diameter for the axis of the two-dimensional subgroup). Thus, the three-dimensional rotation group is a *continuous* symmetry group since it contains the *continuous* two-dimensional rotation group over and over again, with a *continuum* of choices of diameter (rotation axis) direction.

<u>Onboard the Starship E</u>

"Rotate! Rotate!" the captain said.
"Rotate the starship, helm!
Rotate, so that the fate we dread,
will not us overwhelm."
"But captain, you forget one thing,"
first officer points out.
"There is no use in ordering
the ship to turn about.
The laws of nature look the same,
no matter where we face;
in that rotated reference frame,
th' reactions still'l take place."

Slide (translational) symmetry

As we sit on a bench to rest, let us imagine a solid whose atoms are located at the corners of cubes (like children's blocks) stacked end to end in all three directions out to infinity, as shown in Fig. 9.3. The sides of the cubes are empty space, but in the drawing we have sketched them in for clarity.

A repetitive array of atoms or molecules is called a crystal, and for our purpose here, we are considering a cubic array of infinite size. If I slide the (infinite) crystal of our imagination slightly in the direction of one of the sides of its blocks, your

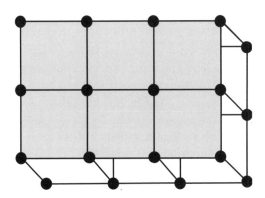

Fig. 9.3. A few of the infinite number of "blocks," whose corners are the actual atoms, stacked together. (In this drawing, the shaded areas are in front.) The lines and shading are included merely to make clear the shapes formed by the array of atoms.

discerning eye will notice that it has been moved. I can move it further in the same direction until each atom is where one of its neighbors had been; the crystal will appear as if it had *not* been moved. I can move it twice, three times, etc., in that direction by the same amount and again reach configurations that appear unchanged. This *changing of position without rotation* is called a (spatial) *translation*. There are an infinite number of such slides or translations that will leave our infinite cubic crystal looking the same; but they are discrete, not continuous — one, two, three, etc. times the length of the basic cube, in one of six possible directions (left, right, in, out, up, or down).

Now, envision instead a *uniform* spread of matter, like ideally smooth peanut butter; there are no atoms in this vision. In our imaginations, this matter also extends to infinity in all directions. Notice that it is symmetric to a translation (slide) of any amount in any direction; the translations form a *continuous* group of operations. This group is the three-dimensional translation symmetry group, sometimes referred to merely as *the translation group*.

<div align="center">

"Translation"

(inspired by the theme of the Volga Boatmen, *DUBINUSHKA*)

"Translate" means move!
"Translate" means move.
Slide along, keeping it so straight.
Thus we translate.
Do not rotate!
Keep a straight groove.
"Translate" means move!

</div>

Reflections

Look! The artist is motioning us to come back to his booth. "I've thought of something else that should *interest* you. My square picture has another discrete symmetry, that is not the same as any of its symmetry rotations. This is not an operation I could ever do to the picture itself, nevertheless it is a symmetry of the picture." Then he holds up a mirror (to the right of the picture). "Now, look at the picture's image in the mirror." The image certainly looks identical to the picture. He then puts numbers (1, 2, 3, 4) in successive corners of the picture, and the result of the reflection becomes apparent. In Fig. 9.4, the picture and its image, produced by the mirror have been drawn flopped over so that we can see the numbers.

Notice that the orientation of the corners in the image is not the same as those in the picture. A clockwise *handedness* or orientation of 1234 in the picture becomes a counter-clockwise handedness in the image. Nevertheless that image is still in the same vertical-horizontal orientation. The artist now places a second mirror to the left of the picture, at a slight angle to the first, so that we can see an image of the first image in it. This is shown in Fig. 9.5. (In the figure, we have omitted the image of the picture in the left mirror, for clarity.)

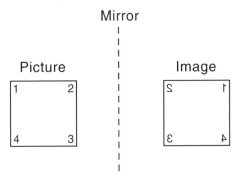

Fig. 9.4. Reflection of the square picture (drawn flopped over so that we can see the numbers), showing the image (also drawn flopped over) in the mirror. Before the numbers are inserted, the image looks identical to the picture. The dashed line shows the edge of the mirror, which faces the picture.

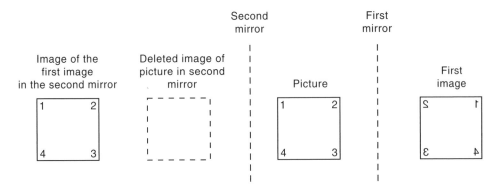

Fig. 9.5. A second mirror has been inserted into the setup of Fig. 9.4. We have deleted the image of the picture, on the left, for clarity. The numbers 1234 are oriented clockwise in the picture, are counter-clockwise in the first image and are again clockwise in the image of that image.

The image of the first image has the *identical* clockwise handedness for its numbers as the picture itself. This shows that two successive identical reflections are equivalent to no reflection at all. The set of operations *reflection, no reflection* is called *the reflection group*.

<u>My Mirror Image</u>

I've a friend who never leaves me.
He's always true and very near.
But he lives 'n a world completely
apart from *this* world — never *here*.

Ever we are separated
my loyal friend, I'll never touch.
Just to watch him, I am fated;
no voice to hear, nor hand to clutch.

And so mocking fate has taunted
the loneliness that aches within.
By this "friend," I'm merely haunted,
as he just mimics grimace 'n grin.

Right is left and left is right,
and clockwise 's counter-clockwise there
in *his* world, that's reached by sight,
as back at me I feel its glare!

The book booth — interchange (commutation) of operations

Now, consider two rotations done in succession. Does it matter which is done first? The answer depends on which rotations are under consideration. If we rotate the square picture one quarter of a turn and then one half a turn, we have a total rotation of three quarters of a turn; if we do the one-half-a-turn rotation first and *then* rotate by one quarter of a turn the result will be the same. This is true for any two rotations of that painting around its central pin; the order that two rotations of the two-dimensional rotation group are performed is immaterial to the final result. The same imperviousness to the order in which they are performed is true for two *translations*. However, rotations in three dimensions do not have this interchange (commutative) property. To see the behavior of three-dimensional rotations, let's go to the book booth and watch the two "magicians" doing their rotations. The magicians have two identical books and start with them in identical orientations: the bindings facing you, the title (T) at the top and S on the left side. (as shown in Fig. 9.6) Magician 1 rotates his book one quarter turn clockwise around your line of sight and then one quarter turn clockwise (when facing left) around a line pointing to your left. Magician 2 performs the same rotations on her book, but in the opposite order.

The final orientations of the two books are completely different, as we see in Fig. 9.6. Reversing the order in which the two rotations are performed has changed the resulting orientation. So we conclude that rotations about different directions do not commute with one another. These magicians have shown us that *the three-dimensional rotation group is not a commutative group*.

Fig. 9.6. Two identical books begin in the same orientation and are then each rotated about the line of sight and the left direction. The order of these rotations is reversed for magician 2 compared to magician 1, which leads to a different final orientation showing the *non-commutativity* of these rotations.

<u>Back Aboard The Starship E</u>

"These operations don't commute,"
first officer points out.
"It's crucial that you institute
a sequence without *doubt*.

Should we first rotate 'round the line
that points t' the nebula,
or do you mean to first align
with th' station's antenna?"

"Yes, yes you're right!", captain replies
"It is not hard to see,
the outcome of these turns relies
on sequence, I agree!

The diff'rence 'tween combined rotations,
with sequences reversed,
is not just your imaginations;
't depends on which comes first."

2. GENERATORS AND CONSERVATION LAWS

Generators of symmetries

The purpose of our journey is certainly not to examine pictures, sculptures, peanut butter or even magicians' books. It is not even to look at crystals. What we are interested in are the symmetries that are inherent in nature itself. Well, what does *that* mean, you are probably wondering. Is it just some mystical blabbering by the author? Who? Me? Nope!

First, we can ask the question: are the laws of nature the same everywhere? More concretely, if we do experiments to discover the laws of nature at one place and do the identical experiments with everything moved (but not rotated) to another place, will we get the same results? In other words are the laws of nature *invariant* to translations (displacements of everything)? In connection with this question, we shall see that such an invariance, in fact, implies the *conservation** of *momentum*. This is one instance of a truly amazing, profound connection between continuous symmetries (here, change of location in space) and conserved quantities (here, momentum).

In order to explore the essence of the argument connecting symmetries and invariances, we must first define the *generator* of a continuous symmetry. We can get a sense of that concept without using mathematics, by considering the translation of a single particle from one point to another. We have seen, in our discussion of the quantum properties of nature, that a particle located in a given region of space is described by a wavefunction that contains a distribution of momenta. Each momentum value corresponds to a simple sinusoidal wave of a different wavelength. If we displace (move) the particle a particular distance, we displace each of these waves by the same amount. Such a shifting of the sinusoidal waves merely changes the phases of the waves at each point in space. (Remember that the phase tells us which part of the wave shape is at that point.) See Fig. 9.7.

Notice that if we displace the wave by a full wavelength, it would have the same phase as it originally had; a displacement by half a wavelength would put it completely out of phase, etc. Thus, the phase change depends on *the fraction of a wavelength that the system has been displaced*. During our discussion of the quantum properties of nature, we have seen that the wavelength of the wavefunction is (inversely) related to the momentum of the particle, so whatever depends on the wavelength of the wavefunction depends on the corresponding momentum of the particle. Therefore the phase change for a particular displacement depends on the momentum. In other words, the *momentum determines the severity of the effect of a translation (displacement)* on the wavefunction. For this reason, *the momenta (in each of the three mutually perpendicular directions of our three-dimensional space)*

*Recall (Chapter 2, Sec. 3) that conservation of momentum means that the total momentum of the universe, or any *isolated* part, remains unchanged in spite of the interactions taking place.

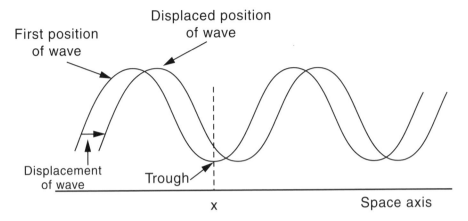

Fig. 9.7. The displacement of a simple sinusoidal wave. A sample point in space has been labeled *x*. Notice that the first position of the wave has a trough at *x*, whereas the displaced position has a higher part of the wave at *x*. Therefore *the phase at x has shifted* from a trough towards a peak.

are called *the generators of translations (in the corresponding directions)*. We have learned that in quantum mechanics the momentum is represented by a mathematical operator, so, strictly speaking, that operator is the generator of translations. Correspondingly, for the translations of a "system" of particles, the operator representing the total momentum of the particles is the generator of translations.

When we discussed the quantum properties of nature, we watched a buoy bobbing up and down as a wave moved by. Imagine that there was a trace-recording machine, with rolling paper (like an EKG-trace recorder), recording the height of that buoy at every instant. The picture shows the height going up and down and up, etc., looking just like the picture of the wave in space. However, here it is giving us the behavior of the wave at a fixed point in space (where the buoy is) as *time* elapses. Suppose we had launched the water wave at a slightly later time. This corresponds to conceptually displacing the wave in time. Comparing it with the undisplaced wave would correspond to Fig. 9.7, but with *x* replaced by *t* and the space axis becoming the time axis, as shown in Fig. 9.8.

The "distance" between peaks on the trace recording is the *time* it takes for one complete oscillation of the buoy; that time is called the *period* (of oscillation). The period plays the same role on this time-trace as the wavelength does on the space picture (Fig. 9.7). In analogy with our result for spatial displacement, we find that for time displacement the phase change depends on what *fraction of a period the time is displaced*. However, the period of oscillation is (inversely) related to the frequency of oscillation* so that *the phase change depends on the frequency*. But, in quantum theory the frequency of the wavefunction is proportional to the energy

*As we have already noted in a footnote to Chapter 4, if the frequency is 2 oscillations per second, then the period (of time) of each oscillation is $\frac{1}{2}$ second (1 *divided* by 2), etc.

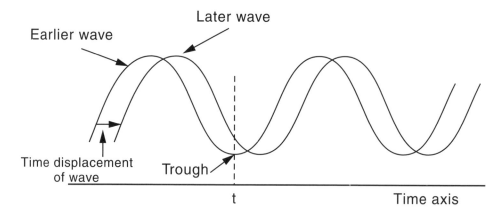

Fig. 9.8. The displacement in time of a simple sinusoidal wave, as shown on a chart recorder
(rolling to the left, so that later times are towards the right). A sample time has been labeled
t. Notice that one position of the wave has a trough at t, whereas the time-displaced wave has a
higher part of the wave at t. Therefore *the phase at* t *has shifted* from a trough towards a peak.

of the particle; consequently, the energy determines the behavior of the system at
a later time. Thus *the energy operator* is the generator of time translations*, which
means that it determines the evolution of the system.

<div align="center">

Space- and Time- Translation Generators

Momentum is the generator — sli-ide
Momentum is the operator — sli-ide
And what of time's elapsing,
with the present instant collapsing?
The energy 's the generator — gli-ide

</div>

Conservation Laws

We have just seen that the operator that generates the response to translational
displacement is the momentum operator. Thus it has two significant properties;
it corresponds to a measurable property (i.e., momentum) of a system and is a
generator of symmetry operations (i.e. space translations) as well. Likewise the
energy operator corresponds to the measurable total energy and is also the generator
of time translation or evolution. We shall see that the dual properties of these
operators lead directly to conservation laws.

Suppose all interactions (i.e., the laws of nature) remain unchanged no matter
how much the system is displaced. First, we note that the interactions contribute
to the energy, so that *unchanged interactions imply that the energy is unchanged*.
This implies that reading the energy *before or after* the space displacement yields
the same result. Since the momentum is the generator of space displacements

*technically known as the *Hamiltonian* operator.

(a.k.a. translations), *the energy and momentum operators commute.* However, the energy operator is the generator of time translations (time evolution); therefore, because momentum commutes with it, the momentum readings are unchanged as time proceeds — *momentum must be conserved.*

Slip, Slide In Vain

Slip me this way, slide me that; you cannot change the rules.
The *laws* of nature do not care how much you shift or fool.
These laws are firm and do not yield, the answer is quite plain;
despite your shifts (so-called *translations*) unchanged *they* remain.

'Tis symmetry, 'tis symmetry – the universal kind.
Let's sing this song of symmetry's translational design.

And there are consequences of location-symmetry.
The corresponding generator is conserved, you see.
That generator is momentum, total, as we've seen.
It is conserved no matter how the particles careen.

Oh symmetry, oh symmetry, your beauty is sublime.
We're dazzled by your sparkling face, yes, each and every time.
Oh symmetry, oh symmetry, your beauty does astound.
We're joyous, for those quantities conserv-ed that abound.

A generalization of our discussion may be obtained by replacing the continuous translational symmetry operations by any other set of continuous operations that do not change the interactions, i.e. that are symmetries of the interactions. The same argument we have just used shows that the physical quantity that corresponds to the generator of that symmetry is conserved since it commutes with the energy operator. Summing up, we see that the generators of *any* continuous symmetry group of the interactions are conserved.

Let us apply this conclusion to the group of rotations. Experiments have shown that the laws of nature are invariant to rotations, and the generators of rotations have been found to be the angular momenta. *Rotational invariance of the laws of nature* consequently leads to the *conservation of angular momentum.*

3. INTERTWINING ROTATIONS

Let's take a closer look at the rotation group. We saw at the fair that the three-dimensional rotation group is a non-commutative group — reversing the order of two rotations that are around *different* directions changes the final result. Since the angular momenta are the generators of rotations, the fact that rotations about different directions do not commute implies that the corresponding angular momentum operators do not commute. As we saw for position and momentum in our discussion of the quantum properties of nature, this means that we cannot measure angular momenta around two different directions simultaneously with absolute

precision. In fact, a system with a definite value of its angular momentum about one direction has a probability distribution of values for angular momenta about other directions.

Remember the two different final orientations of the books in Fig. 9.6? We would have to rotate the book of magician 1 about the *vertical* (and then about our line of sight) to get it in the same orientation as the book of magician 2. But the two rotations done by the magicians were rotations about *horizontal* axes. This shows that horizontal and vertical rotations are not completely unconnected. This *intertwining* of different operations is a property that is ubiquitous in non-commutative groups (like the rotation group).

<div align="center">

i said

so

i love symmetry

symmetry love i

so

said i

</div>

Oh, why should I love symmetry?
Oh, why are we so enthralled,
whenever we find symmetry?
'S some ancient craving recalled?
Well, what is there 'bout symmetry,
that does enchant us so?
Is our delight with symmetry
some echo 'f long ago?

Continuous-group symmetries
are special, that is true.
Their generators we can see
do form conserv-ed crew.
Well, could a symmetry that's terse,
be nature's simple rule
that's fashioning the universe
(yes, even as it cools)?

Simplicity, simplicity,
your beauty does astound,
and all of us are filled with glee
whenever you are found.

Oh, why should I love simplicity?
Oh, why are we so enthralled,
whenever we find simplicity?
'S some ancient craving recalled?
What is there 'bout simplicity,
that does enchant us so?
Is our delight with simplicity
some echo 'f long ago?

ABOUT WHIMSY

Many of the properties of elementary particles that we are about to explore have been given whimsical names. The first property to be named this way (in the 1950s) was "strangeness." This choice of whimsy over presumed profundity avoids the errors of the past, when serious, profoundly meaningful names were chosen like "element", "proton" (from the Greek for first), etc., which were later found to be misnomers.

Whimsy makes no pretense to profundity! Actually, once the gates of whimsy are opened, all fun breaks loose.* Names like "flavor," "color," and several more have this same whimsical nature. However, this playful *whimsy should not misguide us into discounting the significance of these properties of nature. The experimental evidence for strangeness, charm, color, quarks and their properties, gluons, etc.* (all of which we shall explore), *as well as the Standard Model itself, is every bit as solid and rigorous as was the evidence for the establishment of electric charge, the elements, the electron, the proton, the nucleus, the photon, antiparticles, etc.* In fact, the techniques and conceptual bases of recent discoveries were often based upon methods and analyses similar to those of the earlier discoveries. In many cases the earlier investigations served as prototypes for the new analyses.

*a parody on the slang expression of chaotic disaster: "all hell breaks loose."

Now we are walking among the conserved quantities. These lead us directly to the so-called "internal" symmetries. Despite the name, internal symmetries are <u>not</u> those of an internal structure of particles; instead, they are symmetries relating different <u>types</u> of particles to one another. The hypothetical, mathematical space that contains these symmetries is a space of our imagination; there is no way to actually enter this "flavor space." In the course of our journey, we shall see that there are observable physical consequences of these internal, type-of-particle, symmetries. We will begin by looking at isotopic spin as we start our walk along the Trail of Flavors.

Chapter 10

"INTERNAL" SYMMETRIES — ISOTOPIC SPIN

1. CONSERVATION LAWS AND SYMMETRY

There are many different kinds of quantities that are unchanged by particle interactions. We have already met those associated with motion, namely total energy, total momentum and total angular momentum. Their significance goes beyond the fact of their conservation in interactions (including scatterings). We have seen that they are the generators of space-time symmetry groups (translations and rotations), which gives them a much deeper significance than is superficially apparent. Consequently, whenever we find an entity that is conserved, we suspect and *hope* that that conservation law is the result of a *symmetry of nature*.

Let us now envision the "space" of *internal* symmetries, not internal to the particles but rather located within our *imaginations (and our mathematics)*. The symmetry operations here change particle properties and consequently their *types* rather than moving them in spacetime. Let us now proceed to explore these symmetries of particle-type, called *flavor symmetries*, and their associated conserved quantities. Our mental journey takes us to flavor space.

Flavor Space

There is a world within my mind,
fantastic far beyond my dreams,
where particles just change their *kind*
as my thoughts roam to its extremes.

Location'n this "internal" place
corresponds to type and charge.
It has *no time*, and no *real space*.
It has no motion small or large.

2. WELCOME TO FLAVOR SPACE

Isotopic spin — an internal symmetry

First we consider the particles that comprise the nucleus, namely the proton and the neutron, and the strong interaction that is responsible for the nuclear force. We have already mentioned that the nuclear force between proton and proton, proton and neutron, as well as neutron and neutron, are essentially the same. In the realm of the strong interaction, the neutron and proton appear to be two "versions" of the same particle. In fact, their masses differ by less than 1%. Their electromagnetic behaviors are, indeed, completely different — the proton has electric charge, but the neutron is neutral. However, we are now in the *strong* sector of our imaginations. At this point in our journey, before the compositeness of proton and neutron has been surmised, it appears that the slight mass difference between the neutron and the proton will not show up until we travel to the sector of electromagnetic interactions. So, in the strong-interaction part of flavor space, the neutron and proton are considered two different states of the same particle — the *nucleon*. As we have already implied, to distinguish this *symmetry of type* from spacetime symmetries, it is referred to as an *internal symmetry* — a symmetry we can envision in our imaginations, but not in spacetime. This particular internal symmetry is called *isotopic spin symmetry.** It is the first of several flavor symmetries that we shall meet on our journey.

The two states of the nucleon are reminiscent of the two states that exist for electrons in the same spatial state, namely spin up and spin down. We have seen that the spin of a particle is its intrinsic angular momentum, which is understood in terms of the rotation group. The mathematics of rotations, that gave us spin $\frac{1}{2}$ particles, has an analog here in flavor space that applies to the *type* of nucleon. The nucleon is understood to be an isotopic spin $\frac{1}{2}$ particle; when the isotopic spin is up it is a proton and when the isotopic spin is down it is a neutron. And if those particles did not already have their venerable names, we might very well have named them n^+ and n^0, respectively, where n stand for nucleon and the superscript gives the electric charge. We express their relationship to one another by saying that they are in the same *isotopic multiplet*.

Before we proceed further, let us take a closer look at the effect of ordinary space rotations on a spin $\frac{1}{2}$ particle. Consider a state that is spin up along a particular direction. We have already realized, during our contemplations about commutation (after visiting the fair), that a state that is *spin up along another direction will have a mixture of spin up and spin down along the first direction*. Thus, a rotation of the up state to that other direction will produce a state having nonzero probabilities for both spin up and spin down along the original direction: *A rotation of a spin up state will yield a state which is part spin up and part spin down.* By "part" we

*It has nothing to do with nuclear isotopes.

mean that the state has a nonzero *probability* that a measurement will find that result.*

Now isotopic-spin rotations are not rotations in space or in any "internal space," but are instead changes of the *type* of particle — we envision these "rotations" in our minds and can perform them in our mathematics. Since the mathematics is the same as that of space rotations though, a hypothetical *isotopic spin rotation* will change a proton (isotopic spin up) into a state with a part that is neutron (isotopic spin down) and vice versa. This is a description of the mathematical operations involved, although *we can only perform an isotopic rotation in the mathematics and in our imaginations.* Furthermore, we cannot directly verify this mixing because our experiments always find either a proton or a neutron, but not a mixture, since they do have different electric charge. However, the isotopic spin symmetry of the strong interaction does lead to important relationships among the strong interactions of these *different* particles because they are in the *same* isotopic multiplet. Furthermore, when we look at other multiplets, it makes predictions about the *existence* of types of particles based on the discovery of others. We will not look at the detailed predictions of strong-interaction similarities, but rest assured that they have been experimentally verified. However, let us venture further into flavor space to explore the *existence* predictions.

Just as for spin, particles can only have *isotopic* spin (I) values $0, \frac{1}{2}, 1, \frac{3}{2}, 2$, etc. We have just seen that particles with isotopic spin $\frac{1}{2}$ can have their isotopic spin up or down, so that there are two possible states corresponding to two particle types. The "direction" in the isotopic space of our imagination is usually referred to as the "z direction" for short (although it has nothing to do with any direction in actual space); isotopic spin down is referred to as $I_z = -\frac{1}{2}$ and isotopic spin up is the state with $I_z = +\frac{1}{2}$. (I_z is read as "*I*-sub-z" or "I-z.") These comprise an isotopic spin multiplet of two states — a *doublet*. For any value of the isotopic spin (I), the allowed I_z values run in steps of 1 from $-I$ to $+I$: for isotopic spin 1 we have $-1, 0, +1$, which comprise an isotopic spin multiplet of three states — a *triplet*, etc. Remember that isotopic spin refers to *type*, so that *three different types* of particles "dwell" together inside an isotopic triplet. Their I_z values correspond to their electric charges in a manner similar to that of the nucleon. Once we have determined the size of an isotopic multiplet that a known particle is in, we can immediately *predict the existence* of the other particles in that isotopic multiplet even if they haven't been found yet.

*The mathematical structure of the wavefunction is in fact the *sum* of the corresponding two parts.

Isotopic Multiplets

A doublet has two components, yes.
A triplet has three, as you can guess.
A singlet 's alone, with none to mix,
unchanged by the isotopic tricks.

The I-sub-z, it jumps by one
as we so "move," 'til there is none.
Within each multiplet, it "grows."
From minus-I to I it goes.

And if *one* particle does call,
without the use of crystal ball
the other members are surmised
(that's if the multiplet 's well sized).

The particles of an isotopic multiplet behave similarly in the "region" of the strong interaction, so *in the strong-interaction part* of flavor space they are the same *kind* of particle despite the fact that they each have a different electric charge. The isotopic multiplet we have visited so far is the nucleon, with $I_z = -\frac{1}{2}$ corresponding to a neutron (n) and $I_z = +\frac{1}{2}$ corresponding to a proton (p). Next, let us meet the meson that is ubiquitous in strong interaction experiments, namely the pion (π). The pion is an isotopic triplet, π^- has $I_z = -1$, π^0 has $I_z = 0$, π^+ has $I_z = +1$, where the superscript gives the charge of the particle. Here again we have particles whose major differences of behavior (due to their different charges) will first be seen when we venture into the electromagnetic sector.

We can represent an isotopic multiplet by placing the particles along a line according to their I_z values, as shown in Fig. 10.1.

Fig. 10.1. Isotopic multiplets, linking different types of particles into a single category. (a) The nucleons: p is the proton and n is the neutron; (b) The pions: the superscripts show their different charges.

There are many more isotopic multiplets of particles dwelling in flavor space, but we shall not slow our journey down to look at them in detail.

The symmetry group

The group of isotopic spin rotations has a *mathematical* structure similar to that of the (space) rotation group, viz., it has three generators intertwined in exactly

the same way as we saw in our study of noncommuting rotations. And they are conserved (by the strong interactions), just as those of the rotation group are (by all interactions).

By this name, shall we know it

The mathematics of the isotopic symmetry group teaches us that all we really need to know is the behavior of the two states of the isotopic doublet under isotopic rotations, to figure out the behavior of all of the other isotopic multiplets. Consequently, we refer to the group of isotopic rotations as FLAVOR SU(2), where the group name is pronounced as letters and a number "S"- "U"-"2". The "SU" stands for arcane mathematical properties of the group and "2" refers to the two members of the isotopic doublet.

<div align="center">

Isotopic Spin

Isotopic spin 's all right.
For it *is* conserved, despite
the strong interaction's might.
Isotopic spin 's all right.

Isotopic spin 's the theme
'f hadron-class'fication scheme.
FLAVOR SU(2) 's the team.
Isotopic spin 's the theme.

</div>

Mind Travel

And our imaginations soar
beyond all time and space
to realms we can't enter or tour,
that have no size or place.

'Tis here we'll find the secret codes
of nature's structure, grand.
We'll travel on those conjured roads
through 'maginary land.

For here the symmetries we find
provide us with some clues
to how the world became this *kind*,
and where we should now cruise.

We've just met symmetries *internal*
(of type); they've been unfurled.
Those symmetries that form the kernel
of this our arcane world.

They're symmetries whose path, so steep,
on Trail of Flavors winds.
We travel, where we cannot *creep*,
except within our minds.

Now as we venture further into flavor space, we become increasingly aware that the symmetries of flavor space produce observable consequences here on the ground. These consequences appear as quantities conserved by the strong interaction. They are, in fact, generators of the flavor symmetries and are referred to as "flavors." We also come across two numbers that appear to be conserved by all interactions, but seem to have no corresponding underlying symmetries. They are baryon number and lepton number. Their conservation may be merely illusions that will dissipate when our experiments reach unexplored domains or when our understanding becomes more profound.

PARTICLE FLAVORS

We met isotopic spin at the beginning of the Trail of Flavors, and have seen that the value of I_z is related to the charge of a particle in a given isotopic multiplet. I_z is a particle flavor, and we have already had an interesting visit with its symmetry group. Now let us continue our walk to meet some of the other residents.

1. HYPERCHARGE (Y) AND THE GROUP U(1)

The flavor Y — hypercharge

When we look at the isotopic multiplets we have met so far, we notice that I_z is directly related to the charge of the corresponding particle: For the pion, the I_z value is actually the same as the charge; however, this is not true for the nucleon whose charges are *displaced* from the corresponding I_z values by $\frac{1}{2}$. This displacement of the charge values from the I_z values is due to what is called the "hypercharge" — the letter Y is used to designate hypercharge.[*] We will refer to this flavor as "flavor-hypercharge" to distinguish it from its generalization that we shall see appears in the Standard Model (and uses the same letter Y). It has been found that *total flavor-hypercharge is a conserved number in strong interactions.*

[*]The (electric) charge of a particle is the sum $I_z + \frac{1}{2}Y$. The nucleon has hypercharge 1 ($Y = 1$) whereas the pion has hypercharge 0 ($Y = 0$), so that the pion's charges are the *same* as the I_z values.

Flavor-Hypercharge

Charge 's not merely I-sub-z.
Hypercharge 's a part, you see.
Hypercharge 's conserved, they say,
by strong interaction's play.
Hypercharge will not be bowed,
standing tall, as "Y," so proud.
'T generates a symmetry —
FLAVOR U(1), as we'll see.

The group U(1)

In our travels, we have learned that conserved physical quantities like energy, momentum, and angular momentum are the generators of symmetries of the interactions of nature. For the strong interaction, we have already visited with isotopic spin symmetry and the corresponding isotopic spin generators. The conservation of flavor-hypercharge (Y) also leads to an invariance of the strong interaction under a continuous group of operations in *flavor* space generated by Y. This generator appears to be completely independent of the isotopic spin generators, so that we have an independent group with one generator; that group is called FLAVOR U(1). And so, there are other dimensions in flavor space with Y as the generator of the rotations there.

We express the complete independence of the SU(2) of isotopic spin and the U(1) of flavor-hypercharge by saying that the strong interaction is invariant to FLAVOR SU(2) × U(1), where the "×" (read as "cross") means that they are *completely independent* symmetries. Consequently, each isotopic multiplet of hadrons has a definite Y value. Just as the quantity I_z is the associated flavor of the SU(2) part, flavor-hypercharge (Y) is the particle flavor of the U(1). Thus I_z and Y are the flavors here on the ground, which are like mysterious "shadows" of the corresponding flavor-space symmetries.

Antiparticles

The total I_z or Y value of a set of particles is the *sum* of their individual values, just as for total charge; we express this by saying that I_z and Y are *chargelike numbers*. Furthermore, they all have opposite signs for antiparticles compared to their values for the corresponding particles. Consequently, electrically neutral particles *can* be different from their antiparticles. For example, the neutron has $I_z = -\frac{1}{2}$ and $Y = +1$, whereas the antineutron has $I_z = +\frac{1}{2}$ and $Y = -1$; they are quite different even though they both have zero charge. We shall soon explore another chargelike number that is relevant for a discussion of the antineutron.

Chargelike Numbers

Oh, charge is a *number*, for here is the key:
each measurement finds it in units of "*e*."
Yes, quantized in "*e*'s" are the charges we *see*;
the proton's is plus *e*, and, precise-ely,
the charge of an 'lectron is, right, minus *e*.

When particles gather for warm company,
their charges just add up, effective-ely.
Now *Y* fits this profile, yes definitely;
another we've met here is called I-sub-z.
As particle labels, these numbers decree
the "species" of particle that we should see.
They're *chargelike* for they combine additively,
for particles bound and for particles free.
And others are chargelike; additionally,
you'll soon meet conserved numbers S, L, 'n B.

Before you resume your fantastic journey,
please let us now add, most conclusive-ely
(without any doubt, with no hesitancy),
that all anti-particles' numbers must be
the opposite ones, for consist-e-ency.

2. STRANGENESS (S)

Associated production

In the 1950s, many high-energy scattering experiments were done using nucleons
and pions. In some cases, a heavier baryon, a "cousin" to the nucleon, emerged in
place of an incident nucleon. However, the new baryon was *always* accompanied
by a new meson, called a *kaon* (K), as shown in Fig. 11.1, where the heavy baryon
is called the *sigma* (Σ). This phenomenon, the production of a kaon whenever a
heavy baryon was created, was referred to as *associated production*. In a typical
reaction, a pion and nucleon colliding produced in their stead a heavy baryon plus
a *kaon* as final products. However, experiments using the same kinds of incident
particles never yielded a heavy baryon plus a *pion* — there always was a kaon
instead. Actually, we would expect that it is more likely to have a pion than a kaon
produced with the heavy baryon. Nevertheless, nature is saying quite the opposite
— *it is always a kaon, not a pion!*

Strangeness

The production of kaons, but not pions, in these associated-production experiments
was made plausible by the introduction of another number conserved in strong
interactions, which has been named *strangeness*. Indeed, strangeness has been

Remnants of the Fall

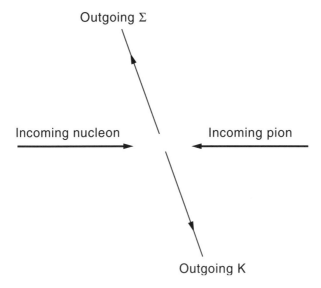

Fig. 11.1. A pion and a nucleon interact producing a heavy baryon (Σ) and a kaon in their stead.

shown to be a property possessed by hadrons (strongly interacting particles) that has integral positive and negative values, just like charge does. In these associated-production experiments, the initial particles have zero strangeness; the heavy baryon produced has a strangeness of -1, and the kaon has a strangeness of $+1$. Thus the final state also has 0 total strangeness, and strangeness is conserved in these associated-production experiments. The production of a pion (whose strangeness is zero) instead of a kaon is forbidden in this strong interaction because it would violate strangeness conservation. (The initial particles have 0 strangeness, but the total strangeness of the final particles would be -1, carried by the Σ.) A consistent set of strangeness values has been determined for the hadrons, such that no violations of strangeness conservation have ever been found in the millions of strong interactions that have been observed.

From a flavor-symmetry point of view, strangeness does not generate a separate group of its own. The flavor-hypercharge symmetry group is enough.*

*This flavor-hypercharge is actually the sum of strangeness and other flavors to be discussed later, and baryon number (to be discussed shortly). Its generalization, extended to leptons, that appears in the Standard Model severs these connections to quark properties.

Strangeness

Strangeness 's not *strange*, but just whimsy at play.
'Tis part 'f flavor-hypercharge, so they do say.
When in the fair Land of Quarks, you will see
it must be taken more seriously.

3. BARYON NUMBER (B)

We have already learned that nucleons and their heavier "cousins" (like the Σ) are baryons. Pions and kaons are mesons. In the nineteenth century, studies of chemical reactions already showed that *the total number of baryons is conserved,* although it was expressed differently* because protons and neutrons had not been discovered yet. Hundreds of baryons have been discovered (since the 1940s) and studies of their many different kinds of reactions found no violation of the *conservation of baryon number* (B). This appears to be a conservation law of *all* of nature, not just of the strong interaction. However, as far as we know, this conservation law is not the result of any symmetry — B is not known to be a generator. This conservation law may be a mirage that will disappear in the light of new experiments.

Baryon number presents us with another difference between the antineutron and the neutron: The neutron has baryon number $+1$ and the antineutron has baryon number -1. In fact, the antineutron is sometimes referred to as an antibaryon.

We have not defined meson number, nor will we ever again mention it. This is because the number of mesons is not even conserved by the strong interaction. In fact, pions are ubiquitous in the products of strong-interaction scatterings, so that a change of the number of mesons is quite usual.

4. LEPTON NUMBER (L)

We have just encountered one important number that does not appear to be related to any symmetry — baryon number, whose conservation appears absolute but mysterious. This conservation law is like a mist, which seems to pervade everywhere but may have no substance to it. There is another number that has the same tenuous existence; it has also been found to be conserved, with no conceptual reason for that conservation. It is *lepton number* (L): The negatively charged leptons (like the electron e^-) and their corresponding neutrinos (like the electron-neutrino: ν_e, pronounced "nu-sub-e" or "nu-e") have lepton number $+1$; the positively charged antileptons (like the positron e^+) and their corresponding antineutrinos have lepton number -1. The electrically neutral neutrinos and corresponding antineutrinos thus differ because they have opposite lepton number.

*in terms of the atomic weight of the chemical elements.

In those decays of nuclei that produce electrons or positrons, commonly known as beta-decays (β-decays), each emerging electron (with $L = +1$) is accompanied by an antineutrino (with $L = -1$), and each emerging positron (with $L = -1$) is accompanied by a neutrino (with $L = +1$), so that the total lepton number remains zero.

No violation of the *conservation of lepton number* has ever been found.

<div align="center">

Numbers – Symmetry?

(inspired by the opening theme of *Carmina Burana* by C. Orff)

Oh, most numbers,
conserved numbers,
generators, seem to be.
Yes, most numbers,
conserved numbers,
generate a symmetry.
But two numbers,
conserved numbers,
leave us wondering openly.
Baryon number,
lepton number,
yes, their source 's a mystery.

</div>

5. RECOLLECTIONS OF CONSERVED NUMBERS AND ASSOCIATED FLAVOR SYMMETRIES

Let us now take a breather, to recall all of the "internal" symmetries and numbers that we have met.

Absolutely conserved numbers?

Baryon number B and lepton number L *appear* to be absolutely conserved. We have found no theoretical reason for their conservation (and, in fact, can construct consistent theories in which these conservation laws are violated.)

Particle Numbers (Flavors) conserved by the strong interaction and their symmetries

We have explored:

- isotopic spin and the corresponding numbers I, from which we determine the multiplet, and
- I_z, which identifies the state in that multiplet of states. The corresponding group is SU(2). I_z *is a chargelike conserved number.*
- flavor-hypercharge Y, which, when combined with I_z, tells us the electric charge of the corresponding particle. The corresponding group is U(1). Y *is a chargelike conserved number.*

- strangeness S, which is related to the flavor-hypercharge of the particle. There is no separate group corresponding to this number; the U(1) of flavor-hypercharge already accomplishes the task. Nevertheless, *S is a chargelike conserved number.*

An absolutely conserved number

Although the flavors discussed above are conserved only by the strong interactions, a particular combination of I_z and Y is conserved by all the interactions. It is the total charge.* We shall investigate this feature further along in our journey.

With the gates of whimsy already opened, it is not surprising that the isotopic spin symmetry and flavor-hypercharge (which subsumes strangeness) symmetry are referred to as *flavor symmetries*. This name was chosen to distinguish them from the *color* symmetry we shall find in the QCD mountains of the Standard Model.

*The charge (Q) is the sum: $Q = I_z + \frac{1}{2}Y$.

Flavors and Friends

Those symmetries that we now greet,
 they play in our mind's band.
The *flavor* numbers hear their beat
 and sing across our land.

There's isotopic spin, of course,
 now hypercharge* resounds.
And flavor symmetry 's the source
 of their sonorous sounds.

And strangeness now pipes up, so loud,
 while singing 'ts plaintive line.
As part of hypercharge, 'tis proud
 to sing with a band this fine.

These flavors are always conserved
 by that strong interaction.
Their voices show it is deserved,
 such beauty, skill, and action.

Their songs and echoes fill the air;
 our particles, they seize.
And our world sways, both here and there,
 to flavor melodies.

All sing the song of symmetry
 with voices sweet and clear,
with covered tones of such beauty,
 that res'nances appear.

Oh symmetry, oh symmetry
 your beauty does astound,
and all of us are filled with glee
 whenever you are found.

But flavor numbers cannot play
 outside the *strong* region.
Weak interactions cause decay,
 cons'rvation is undone.

And yet one combination *soars*
 conserved without a doubt.
'Tis charge (electric) thundering o'er
 them all, with glor(i)ous shout.

Right there up front, in open sight,
 the baryon number stands.
Mysterious, so tall and bright
 conserved in *all* the lands.

And lepton number 's there aglow,
 as proud as it can be.
No symmetry commands although
 conserved consistently.

Both baryon 'nd lepton amount,
 conserved by all we see;
these numbers sing duets that *count*
 and charm both you and me.

We know no symmetries that reign,
 that to *these* numbers call.
But so far we cannot attain
 a change of these *at all!*

Perhaps these two are vivid mists,
 that all around pervade,
and when the "light" shines on their lists
 their conservations fade.

*We refer here to flavor-hypercharge, not its generalization that appears in the Standard Model.

As we continue our trek along the Trail of Flavors, we reach a larger flavor symmetry. This symmetry provides the structures that serve as a periodic scheme for the strongly interacting elementary particles (the hadrons), and also points the way up to the Land of Quarks. From our vantage point in the foothills of the Land of Quarks, we are struck by the similarity in the generation-structure of the quarks and that of the leptons. We also begin to realize how the flavor symmetries flow from the flavors of the quarks themselves.

A LARGER FLAVOR SYMMETRY — ENTRANCE TO THE LAND OF QUARKS

1. A HISTORICAL NOTE: THE PERIODIC SCHEME OF THE ELEMENTS

In the middle of the nineteenth century, two scientists (Mendeleyev in Russia and Meyer in Germany) independently showed that the chemical elements have a periodic repetition of their properties as the atomic number increases. Their success depended upon the recognition that in order to construct a meaningful table of elements, some blank spaces for elements that had not yet been found were required. Thus they predicted the *existence* of new elements which were indeed found thereafter. The table we refer to here is the periodic table of elements that hangs on the walls of almost all elementary chemistry classrooms. The reason for this periodic chemical behavior was found in quantum mechanics; it required the discovery of the electron and the nucleus — the underlying *constituents* of the atoms of those elements.

2. FLAVOR SU(3) AND THE PERIODIC SCHEME OF THE HADRONS

The structure of the group

Let us pause to consolidate what we have seen about the structure of symmetry groups. We saw at the fair that the group of three-dimensional rotations in ordinary space has embedded within it groups of two-dimensional rotations.* Furthermore, rotations about two different directions do not commute but instead are related to rotations about other directions. (Recall that two rotations about different *horizontal* axes were related by a rotation about the *vertical* axis.) This type of (completely) intertwined group is called a "simple group." We have met a couple of simple groups

*and that there is a continuum of choices of axes (diameters) for these two-dimensional rotations.

already in our travels, namely the rotation group (including spin) and the isotopic spin group. Those two groups both correspond to the mathematical group SU(2). On the other hand, the internal strong-interaction flavor symmetry group we have visited is called SU(2) × U(1), where the SU(2) refers to isotopic spin (I) rotations, the U(1) refers to rotations in flavor-hypercharge (Y) space, and the "×" (read as "cross") means that these two *separate* groups are merely "pasted" together. So groups with this × (cross) structure are definitely not the *simple* groups that would fulfill our desire for unification! Indeed, the goal of finding a *unified* theory of interactions, which has occupied the attention of scientists for centuries, would require a *simple* symmetry group, rather than two that are pasted together in the "cross" relation described above.

Let us look at the simple group that succeeds in incorporating these SU(2) and U(1) flavor symmetries in an intertwined manner. It is called FLAVOR SU(3).* We shall see that FLAVOR SU(3) succeeds in providing an elegant means of classifying the strongly-interacting elementary particles. Furthermore, *careful consideration of this group led to the discovery of quarks.* Thus concepts and their relationships within an abstract and arcane area of mathematics amazingly lead us to an understanding of the composite nature of the particles we see.

<div align="center">

SU(2) × U(1)

S U 2

cross U 1

just a mindless combination.

For the "cross"

stands right there,

their independence to declare.

S U 3

we can see

contains those groups with subtlety.

S U 3

sleek design!

Its generators intertwine.

</div>

The periodic scheme of the hadrons

In our discussion of the *possible* multiplets of isotopic spin (SU(2)), we saw that each multiplet can be identified by the number of states it contains — singlet, doublet, triplet, etc. Likewise, the multiplets of FLAVOR SU(3) are named by the number of states *they* contain, using either number or name. Those of particular interest are the **1** (also called an SU(3)-singlet), the **3** (also called an SU(3) triplet), the

*FLAVOR SU(3) contains the mathematical structure SU(2) × U(1) in the continuum of ways that the two-dimensional rotation group is contained in the three-dimensional rotation group. FLAVOR SU(2) × U(1) is one of these embeddings, with the I_z and Y "directions" being unique inside the group of FLAVOR SU(3) operations.

$\bar{\mathbf{3}}$ (read as "three-bar" and also called an SU(3) antitriplet), the $\mathbf{8}$ (also called an SU(3) octet), and the $\mathbf{10}$ (also called an SU(3) decuplet). Notice that there are two different multiplets with three states (the $\mathbf{3}$ and the $\bar{\mathbf{3}}$).

The *several hundred types of hadrons* (strongly interacting particles) that we have passed (but not met) along our way, have been shown to fall clearly into SU(3) $\mathbf{1}$s, $\mathbf{8}$s, or $\mathbf{10}$s. We are about to visit Fig. 12.1, which contains examples of the lightest FLAVOR-SU(3) multiplets. Whereas the isotopic multiplets were all one-dimensional (recall the I_z lines in Fig. 10.1), the multiplets here have two dimensions — one for I_z and one for Y.

In Fig. 12.1(a), we see the (two-dimensional) baryon octet ($\mathbf{8}$) that contains the baryons we have already met, namely the nucleon (the **n**eutron and the **p**roton) at the top and the Σs (sigmas) in the middle. Figure 12.1(b) shows the lightest meson octet, which contains our friends the π (pion) and K (kaon), as well as the antikaon ($\mathbf{K^-}$ and $\mathbf{\bar{K}^0}$). We need not meet any of the others in these octets, but should be aware that every point in each of them represents an *observed* particle. It is the *intertwined* nature of SU(3) that has linked *different* I_z lines into one multiplet.

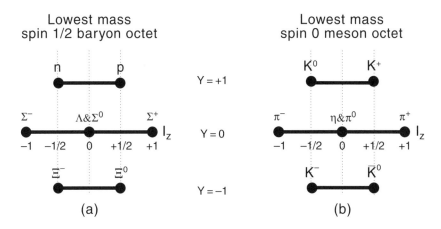

Fig. 12.1. The lightest SU(3) octets, showing the I_z and Y values of their respective particles; (a) shows the octet of spin $\frac{1}{2}$ baryons that contains the nucleon states (**n**eutron and **p**roton), (b) shows the octet of spin 0 mesons that contains the pion and kaon states (as well as antikaons $\mathbf{K^-}$ and $\mathbf{\bar{K}^0}$). The thick solid (horizontal) lines connect particles of the same isotopic multiplet.*

The lightest spin $\frac{3}{2}$ particles that occur form an SU(3) decuplet ($\mathbf{10}$), filling out the pattern expected from SU(3) considerations with the correct I, I_z, and Y values, as shown in Fig. 12.2. When FLAVOR SU(3) was proposed, all of the particles in this decuplet had been found *except* for the Ω^- (omega minus) sitting at

*The names of the other particles in the baryon octet are: Λ ("lambda"), an isotopic singlet; and Ξ (called "the cascade," because they decay rapidly to Σs or Λs that then decay to nucleons — a cascade of decays). The extra particle in the meson octet, η ("eta") is an isotopic singlet.

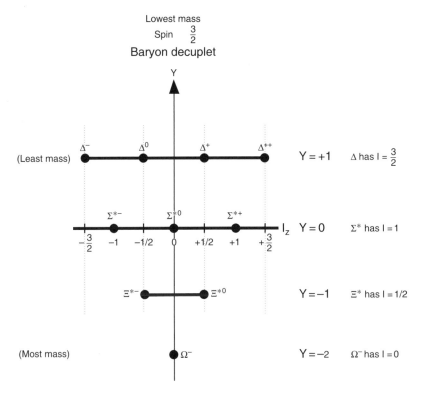

Fig. 12.2. The lightest $SU(3)$ decuplet of (spin $\frac{3}{2}$) baryons, showing the I_z and Y values of their respective particles. The existence of the Ω^- was predicted by the $SU(3)$ classification scheme and subsequently found.*

the bottom of the diagram. Just as was the case for the periodic table of elements, where missing elements were predicted to exist by *that* scheme, the existence of the Ω^- was predicted *before* it was found (right at the mass predicted for it).

The multiplets of Figs. 12.1 and 12.2 form the *hadron classification scheme* that repeats over and over again as we go higher in mass. All hadrons belong to such multiplets; furthermore, the $SU(3)$ group implied by the existence of these multiplets will lead us directly the Land of Quarks!

The clues that led to the discovery of the isotopic spin $SU(2)$ symmetry group included the fact that particles in the same isotopic multiplet have the same strong interactions and essentially the same mass. However, the different isotopic multiplets contained in *one* FLAVOR-$SU(3)$ multiplet have *different* masses with differences that are much larger than the slight variations within each isotopic multiplet. For example, as we move down in Fig. 12.2 the mass increases as indicated there. Nevertheless, these (least massive) spin $\frac{3}{2}$ particles do form a decuplet of $SU(3)$,

The particle names, from top to bottom, are Δ ("delta"), Σ^ ("sigma star"), Ξ^* ("cascade star") and Ω^- ("omega" or "omega minus").

just as the least massive spin $\frac{1}{2}$ baryons and the least massive spin 0 mesons form octets with the expected I, I_z, and Y values (Fig. 12.1). Furthermore, *all* of the heavier hadrons (strongly interacting particles — both baryons and mesons) also form FLAVOR SU(3) multiplets.

Now, let us recall that even the isotopic spin symmetry that we have so readily embraced is not a symmetry of the *world*, but is a symmetry only of the strong interaction. It is broken by the electromagnetic interaction, which treats each member of an isotopic multiplet differently. For example, the proton interacts quite differently than the neutron does because the proton has charge and the neutron does not; likewise, in *every* isotopic multiplet, each member has a different amount of charge. And so, we are inspired to guess that analogously FLAVOR SU(3) symmetry has some significance, although it is a broken symmetry even of the strong interaction.

<div align="center">

Hadron Classification Scheme

Isotopic Spin 's the seam
'f hadron class'fication scheme.
Hypercharge is right there too,
joining in the flavor stew.

They form FLAVOR SU(3),
well-known *broken* symmetry.
SU(3) has multiplets;
hadrons group into these sets.

</div>

Quarks

The triplet (**3**) and antitriplet ($\bar{\mathbf{3}}$) FLAVOR-SU(3) multiplets do not correspond to any observed hadrons; however, the mathematics shows that particles which belong to any of the larger multiplets might very well be composites made of triplet and/or antitriplet particles. Such compositeness would explain, in a simple way, why the hadrons fit into SU(3) multiplets. So, we might guess that such triplet and antitriplet particles actually exist and that they are the constituents of the hadrons. These have been named "quarks"[*] and their corresponding antiparticles are called "antiquarks." This quark hypothesis neatly explains why FLAVOR SU(3) multiplets are formed, namely because the particles in those multiplets are made from *three* different kinds of fundamental quark constituents. Furthermore, we shall see that the broken nature of this flavor symmetry follows simply from the fact that those three quarks do not all have the same mass.

But there is a puzzling feature of the collection of *observed* strongly-interacting particles. No particles have ever been found forming any other SU(3) multiplets besides **1**'s, **8**'s, and **10**'s. Even the quarks themselves (which are flavor triplets)

[*] a word taken from *Finnegan's Wake* by James Joyce — "Three quarks for Muster Mark." The name was originally selected for its sound, and later the spelling was chosen from this literary source.

have *not* been found directly. However, the mathematics shows that combinations of three quarks can only be **1**'s, **8**'s, and **10**'s, just those *seen to occur* for baryons; likewise, a quark and an antiquark can combine to produce only **1**'s and **8**'s, just those *found experimentally* for mesons. Thus, the mathematics, backed by strong experimental evidence, beckons us into the Land of Quarks.

3. ENTERING THE LAND OF QUARKS

Introducing the residents

Now we are high into the foothills of the mountains. The dialect is different here, and the words "isotopic spin" are no longer used, replaced by a *more general* concept called "isospin." Here we find the *sources* of the flavor symmetries we encountered below. In fact, the word "flavor" will henceforth be used for the quarks themselves, to denote their different types. The strongly-interacting particles (hadrons) we have met so far are composites of three flavors of quarks: *up, down,* and *strange* (and their corresponding antiquarks). The old isotopic spin refers only to the *up* and *down* quarks, which make up the protons and neutrons in the matter all around us. (We have only become acquainted with "strange" matter by studying high-energy interactions.) The (almost) unbroken property of isotopic spin under the strong interactions arises from the near equality of the *up* and *down* masses. The larger mass differences in the multiplets of FLAVOR SU(3) occur because of the larger mass of the strange quark. In Fig. 12.2, the particles on the top line have no *strange* quarks and so are the least massive, those on the second line have one *strange* quark, those on the third line have two *strange* quarks, and the bottom one (the Ω^-) is made of three *strange* quarks. (There are similar mass differences that exist in Fig. 12.1.)

There are three *heavier* flavors of quarks that we have not met yet, which were discovered after the existence of quarks was surmised from FLAVOR SU(3). Altogether, six flavors of quarks exist. Furthermore, they each come in three possible *colors* (more whimsy), a concept that will be explored later in our journey. Now that comes to 18 different kinds of quarks, and we must not forget their corresponding antiparticles; altogether there are 36 different types of these fundamental constituents of the hadrons.

Although the quarks are the constituents of the hadrons, *we cannot see quarks directly*. Many intensive searches for quarks have been undertaken — in tons of sea water, tons of earth, moon dust, etc. — but no evidence of their existence was ever found. This contributed to the doubts that people had about the quark theory during the 1960s. However, we shall soon see that the strong interactions can be explained by the theory of quantum chromodynamics (QCD), which also shows why the quarks were never found. We shall see, when we reach QCD, that the quarks "live" deep inside gluon bags. There is not enough energy in the entire world to

pull them out, so their colors remain *hidden* (as do the colors of the gluons, as well). This *confinement of color is absolute*, as far as anyone has been able to determine. Nevertheless, a combination of theoretical and experimental *ingenuity* has shown definitively that there exist three isodoublets of quarks. (Here we are talking about isospin!):*

$$\begin{pmatrix} up \\ down \end{pmatrix}, \begin{pmatrix} charm \\ strange \end{pmatrix}, \begin{pmatrix} top \\ bottom \end{pmatrix}.$$

Each of these six types of quarks is considered a different quark flavor. And each doublet is a different *generation*:

$$1^{st} \; generation: \begin{pmatrix} up \\ down \end{pmatrix},$$

$$2^{nd} \; generation: \begin{pmatrix} charm \\ strange \end{pmatrix},$$

$$3^{rd} \; generation: \begin{pmatrix} top \\ bottom \end{pmatrix}.$$

And Don't Forget Their Antiparticles

We now *know*
three generations of quarks.
Even though
we cannot free them like corks.
The first one 's
in the stuff all around us.
The next ones
that we've found, don't confound us.
There are three
generations it is true,
and every
generation contains two.
That'd mean
that there are six kinds of quarks.
But eighteen
is the *colorful* answer, of course.

In the spirit of whimsy, which created the names "strange" and "charm," the third generation of quarks is *sometimes* referred to as:

$$3^{rd} \; generation: \begin{pmatrix} truth \\ beauty \end{pmatrix}.$$

*The *upper entry* in each of these doublets has $I_z = +\frac{1}{2}$ and electric charge $+2/3 \, e$, and the *lower entry* has $I_z = -\frac{1}{2}$ and electric charge $-1/3 \, e$. Note that their electric charges are fractions rather than integers (times e) — $+2/3 \, e$ and $-1/3 \, e$ for quarks, and their negatives for antiquarks. No fractionally charged particles have ever been found.

<u>Short Ode to John Keats</u>

Beauty ain't truth;
truth ain't beauty.
But they do form a pair.
These are the quarks of the third generation
whose existence is certainly rare.

The masses of the quarks in each generation are larger than the masses of those in lower generations. The FLAVOR-SU(3)-multiplet structures of the hadrons we have encountered so far result from the fact that they are composed of the three lightest quarks: *up, down,* and *strange.* (The *charm* quark is effectively three times as heavy as the *strange* quark.) In fact, even the *strange* quark is sufficiently heavier than the *up* and *down* quarks so that the masses within each individual FLAVOR-SU(3) multiplet are considerably spread (those with *strange* quarks or antiquarks are heavier), as we saw in Fig. 12.2. The *charm, bottom* and *top* (or *truth* and *beauty*) quarks are so much heavier than those lighter three that it took extraordinary ingenuity, skill, and imagination to ascertain their existence — no symmetry-group aid in sight.

The concept of quarks was created in 1963 but was not *embraced* by the physics community until convincing solid *experimental* evidence was found, beginning in 1974 with the discovery of the *charm* quark. This is just as should be expected of a modern science!

Quark Flavors

The *up* and *down,* they have no *class*;
they're plain and very light.
But do the heavier quarks surpass
in bringing us delight?

There's *strange* and *charm,* with names that please,
but names do not portray
the character or properties,
although they're fun to *say.*

Now *top* and *bottom* sound so bland,
but *beauty* does astound.
And *truth* rings out across the land,
with rich and glorious sound.

But, in truth, the quark that's *truth*
as *top* is known these days.
Extremely heavy, but forsooth
it rapidly decays.

And how does *beauty* come to be
just *bottom* plain and square.
Yes what's become of whimsy's glee
with words that tease and dare.

'Tis heresy to e'en propose
new names for *up* and *down.*
For surely, most would those oppose
with sneer and scornful frown.

Lepton generations

Now, from this vantage point high in the foothills of the Land of Quarks, we look down at our path below. There we see that the leptons form isodoublets* too:

*The upper entry in each of these doublets has $I_z = +\frac{1}{2}$ and the lower entry has $I_z = -\frac{1}{2}$, just as do the quarks.

$$1^{st} \text{ generation:} \begin{pmatrix} \text{electron–neutrino} \\ \text{electron} \end{pmatrix},$$

$$2^{nd} \text{ generation:} \begin{pmatrix} \text{mu(on)–neutrino} \\ \text{mu(on)} \end{pmatrix},$$

$$3^{rd} \text{ generation:} \begin{pmatrix} \text{tau–neutrino} \\ \text{tau} \end{pmatrix}.$$

Here we see six lepton flavors. The mathematical consistency of the Standard Model has, in fact, been shown to *require* that the number of quark generations must match the number of lepton generations. Again, we find that abstract, arcane mathematics appears to have significance in the physical world!

Experiments in particle physics as well as calculations based on astrophysical observations, two very diverse physical studies, have established the fact that there are *only* three generations. Other kinds of particles may exist, but they cannot be simply more generations containing leptons with massless or nearly massless neutrinos.

Symmetry Song in Counterpoint

echo:	*Let's sing a song of symmetry*
Chorus 2:	Climbing, climbing
echo:	*that echoes in our minds,*
Chorus 2:	higher, higher.
echo:	*and leads to thoughts of such beauty*
Chorus 2:	ever nigher
echo:	*about the world's design.*
Chorus 2:	t' concepts chiming.
echo:	*about the world's design.*
Chorus 2:	Symmetry \cdots, symmetry \cdots, symmetry \cdots, symm…

Part III

ON TO THE STANDARD MODEL AND BEYOND

Now we are entering the heart of symmetry land. We have ascended high above the flavor symmetries that flowed from the existence of three quark flavors whose masses were not extremely different from one another. Those flavor symmetries led to our awareness of the quarks' existence (and subsequently, particles containing heavier quarks have been observed). As we enter the realm of "gauge theories," we will learn the remarkable fact that the interactions themselves are each the result of a very different, special kind of symmetry known as a "gauge symmetry." Symmetries of this kind are at the heart of the theories that account for the world as we know it, so we now pause to learn their general structure and properties.

Chapter 13

ABOUT GAUGE THEORIES

1. THEIR "LOCAL" PROPERTY

When we first met symmetries, at the symmetry art fair, we saw an artist rotate his strangely hung paintings about their central pins. In each case, the entire painting was rotated by the same amount. If he were to try to rotate the inner circle of his painting but hold the outer ring firmly in place (see Fig. 13.1), the painting would tear to shreds.

Fig. 13.1. The circular painting, showing the inner circle and the outer ring.

Rotating the outer ring, but by a different amount than the inner circle would also be inadvisable. In fact, rotating *any* part by a different amount than *any other* part would result in the same artistic disaster.

But suppose the artist had told us that he *could* rotate the inner circle of his circular painting without rotating the outer ring at all. We would know that those two parts of the picture were not directly attached to each other. Naturally, we would want to take a look at the back of the painting.

Now let us jump to the concept of internal symmetries. Imagine that we have an *internal* symmetry (a symmetry-of-type) in a *new* conceptual space — not in the flavor space we already understand in terms of the quark flavors. We have seen that a symmetry-of-type exists within the mathematics and within our minds, but is not

part of the space we live in. Now imagine that at each point in physical space, the space we *do* live in, we perform a different (arbitrarily chosen) *internal* "rotation", i.e., we perform transformations, in the type-of-particle space of our imagination, that are different at each point in physical space. Such operations are referred to as "local" transformations because *they are different at each point in physical space*. In our considerations of *flavor* space, the flavor "rotations" were the same at each point in physical space, so that we didn't even mention "physical space." Those kinds of operations are sometimes called *global* transformations to distinguish them from the local transformations that we are now considering.

You might guess that something should "tear" when we perform local internal transformations. The word "tear" is in quotes because an *internal* space is very different from physical space. However, the analogy is qualitatively correct; such transformations in the internal space cause a violent change, unless there is something else present. Just as the circular picture would tear without some unusual connections in back, the internal space of our imagination (and the mathematics) requires special structures and connections, in order for the theory to be unchanged by these *local* transformations.

2. GAUGE BOSONS AND THE INTERACTIONS

In a mathematically arcane manner, the gauge-boson fields serve as the hidden "connections"[*] between space-time points, to prevent the "distortions" that would result from local transformations in their absence. Yes, the mathematics shows without equivocation that symmetry under *local* (different at different physical places) internal transformations *requires the existence of a quantum field, whose quantum particles are massless and spin 1*[†] (one type for each generator of the symmetry group) to "absorb" the resulting "stress." Those particles of the gauge field are the gauge bosons we have met before. Furthermore, this *local symmetry dictates exactly how* those gauge bosons interact with the particles of matter. A theory with this kind of structure is called a *gauge theory*.

<div align="center">Gauge Symmetries Interview</div>

Q. Can we "rotate" *internal* space
a *different* amount at each spot?
Real space would then be torn, disgraced,
destroying the spacetime we've got!

A. Gauge fields are the "connections" there
(those fields that are quantized — spin one);
they grab each particle with care.
That's how *those* rotations are done.

[*]Mathematicians actually refer to these fields as "the connections."

[†]except for gravity, where the theory is more complicated and the resulting *graviton* has spin 2.

Local Internal Symmetries

Now let us get this straight and clear,
 our minds must be allowed
to 'nvision spaces, never *here*
 yet real and just as proud.

Dimensions in these spaces can
 be "colors," isospin,
or hypercharge along the span;
 they are all genuine!

And even more demanding yet,
 we must be free to twist
those spaces 'round, without a fret,
 secure they won't resist.

No mysticism here gives pause,
 no vagueness to discern,
as at each point in space we cause
 the "inner" space to turn.

For even though those turns may be
 quite different at each place,
the gauge bosons, with dignity,
 absorb the twists with grace.

Just as these *local* symmetries
 depend on *them* for aid,
their *interaction*-plans' decrees
 are consequently made.

The electromagnetic interaction, which is an apparent part of our lives each day from the time we shut our alarm clocks and open our eyes to the *light*, can be understood to be the result of such a local (gauge) symmetry; the *photon* is the corresponding gauge particle. All of the electromagnetic interactions — the electric power that lights our world and is necessary for the operation of those alarm clocks, radios and televisions, etc.; the interactions of charged particles and even light itself — are completely understood in terms of the gauge theory we call *quantum electrodynamics* (QED). Actually, both QED and the classical theory upon which it is based were developed piece by piece, detail by detail from the symbiotic interplay of fundamental experimental results and conceptual mathematical deliberations over a period of about two centuries. It was realized in the middle of the twentieth century that that beautiful and powerful theory, experimentally verified to extremely high precision, results from the apparently simple requirement of invariance to a *local* internal symmetry. The electromagnetic interaction is completely determined by that requirement, except for its overall strength, i.e., the size of the fundamental *electric charge*. We find that value, along with the masses, from experiment.

Q E D - Gauge Symmetry

Oh Q E D, Oh Q E D,
can it be true, old friend
that you're a masked gauge symmetry,
we didn't comprehend?

And is our pal, the photon fleet,
a "c'nnecting" gauge-field ray
that reaches all, as charge it greets
and grabs that *special* way?

The type of internal symmetry that produces all of quantum electrodynamics is a U(1) symmetry (with only one generator) — *mathematically* the same as the (global) U(1) we met when we visited with the flavor-hypercharge symmetry, but referring to completely different interactions. So, we can now say that quantum electrodynamics is a U(1) gauge theory with the photon as its massless gauge particle.

Before we entered the realm of gauge theories, symmetries aided us in understanding the *existence* of particles that were partners (in their multiplets) of the particles we already knew, and also accounted for the *relationships* of their respective interactions. Continuous symmetries provided us with a rationale for the *conservation* of important observables. The gauge symmetries have such amazing power that they also *determine* the interaction itself along with the *existence and properties* of the gauge particles that transmit the interaction! As we climb higher into the mountains, we will find that the strong and the weak interactions are also described by gauge symmetries and we will see an amazing connection between the weak and the electromagnetic.

At this point in our journey, we find the *simplicity* with which all gauge symmetries *completely* determine the form of their interactions to be so intriguing, that we enthusiastically embrace the notion that a gauge symmetry underlies all of the structure and interactions in the universe.

The Gauge-Gods Dynasty

Out there "above" our spacetime shell
the Gauge-God Kingdom lies.
Within its glor(i)ous realm there dwell
those mighty symmetries.

That fabled land of *type* is theirs.
"Internal" is its name.
Though not in *space*, we sing sweet airs
its glory to proclaim.

The Gauge Gods reach each spacetime spot
with all-pervading fields.
They grab the "handles" of each dot
and thus gauge power wield.

Yes charge and hypercharge they seek;
when "color" feels their might,
bold color-havoc they then wreak,
though they are out of sight.

Gauge symmetries, gauge symmetries
your beauty is sublime.
Your detailed laws, all your decrees
reign over all spacetime.

Gauge symmetries, gauge symmetries,
your dictates are supreme.
Throughout we feel your majesty
and sway 't your cosmic themes.

3. *THE IDEAL* VACUUM AND RENORMALIZATION

The words "THE IDEAL" are in italics in the title of this section for a reason. A vacuum is ordinarily achieved by sucking the air out of a container; the fewer the molecules left, the closer the region inside is to a true vacuum. In fact, that's how the "vacuum cleaner" got its name; it sucks air in (although not attempting to produce a vacuum). "*The ideal* vacuum" is something quite different. It is a hypothetical state of the world, not as we know it but as it would be with nothing in it at all — no particles of matter, no particles of the interaction fields, nothing at all! This state is indeed an idealization,* but of great importance to our understanding of quantum field theories. How do quantum field theory calculations describe this *nothingness*? It should be the easiest to describe of all conceivable states. However, the mathematical structures of our theories show virtual particles being produced (in pairs) and then annihilated at random places and times. This phenomenon is called *vacuum polarization*.

*It is not the vacuum upon which our universe is based, but that cosmological tale is not on our itinerary.

Attempts to "find" virtual particles will fail unless they are given enough energy and/or momentum to become real. However, we know that an *undisturbed completely empty* space would remain that way everywhere indefinitely, so that we must *swallow* all this extra stuff (virtual pairs floating in and out of existence) in the vacuum state as *artifacts* of our mathematics. Whether there is any *reality* to these mathematical creatures (the virtual particles) is a metaphysical question; however, considering them real is the simplest way to understand the content of our quantum field theories, as we have already seen during our visit with the gauge particles. They provide picturesque descriptions allowing us to visualize what is likely to occur, and they certainly are real *mathematical* (and diagrammatic) creatures of our quantum field theory calculations. When we use our quantum field theories to calculate interactions among particles, these virtual particles must always be taken into consideration, but we must "ignore" (in a very precise and logical manner) the same portion of their shenanigans that was ignored when we discussed the vacuum state. In a way, the vacuum state is operating as a vacuum cleaner for our calculation, removing the most "dirty" parts! This arcane treatment of the virtual particles and their *vacuum polarization* artifacts is called *renormalization*.*

<div align="center">The Virtuals</div>

Soprano solo $(mf)^\dagger$:	Now take your virtuals my son,
Chorus $(mp)^\dagger$:	*Those particles you cannot see;*
Sop. solo	you know you cannot go without.
Chorus	*they're virtual, they're virtual.*
Sop. solo	They're dangerous when you've begun,
Chorus	*Yet they're as robust as can be,*
Sop. solo	but you'll renormalize no doubt.
Chorus	*but virtual, yes virtual.*
Sop. solo	You are so handsome and so true,
Chorus	*Mysterious, yet always near*
Sop. solo	as long as virtuals, you use.
Chorus	*still virtual, yes virtual.*
Sop. solo	You must take them before you're through
Chorus	*Tell us the meaning, make it clear,*
Sop. solo	if not, all power you will lose.
Chorus	*of "virtual", this "virtual."*
Sop. solo	My son, so glorious and proud,
Chorus	*Without them, we can't understand*
Sop. solo	you know, my dear, what is at stake.
Chorus	*(though virtual, just virtual);*
Sop. solo	You must stand always straight, unbowed,
Chorus	*the grand design of nature's plan*
Sop. solo	your grand domain please don't forsake.
Chorus	*needs virtuals, the virtual.*

*Some people like to speak of a part of the shifts as "zero-point" energy, but this is not relevant here.
†*mf* is short for *mezzo forte*, meaning medium loud; *mp* is short for *mezzo piano*, meaning medium soft.

In *renormalizable theories*, the ignored parts of the calculations merely shift the values of charge and mass of the particles from whatever symbolic values we put into the initial step; they do *nothing else*. (Those initial values are *symbolic* because there is no way to measure them separately.) Since we know the *actual* values of the masses and charges of the particles from direct measurements, we neatly tuck away all of these presumed artifacts of our calculational techniques, the vacuum polarization effects combined with the initial symbolic values, and replace them by the observed charges and masses. After all, they *must* add together to give those (already-measured) values. And after all of this is done the symmetry is still unbroken, which ensures that *the gauge particles are massless* even after renormalization. From there on, we calculate the results of scattering reactions and other interaction events in terms of those previously measured values of the masses and charges. Actually, when we probe very close to the particles by means of high-momentum scatterings, the *effective* charge changes with the value of the momentum used. The nature of this observed behavior of effective charge is clearly *understood* from our renormalization considerations.

The dirty little secret that I carefully avoided telling you until now is that the parts of our calculations that are incorporated into charge and mass renormalization actually shift the values of charge and mass by *infinite* amounts. WOW! — NOT TO WORRY! After all of the sweeping is done, and we use the previously measured values for the charge and mass, the predictions for interactions are amazingly successful. All of the predictions made by quantum electrodynamics for many different kinds of electromagnetic interactions have been verified by experiment to *accuracies of millionths of a percent. Quantum electrodynamics is the most accurately verified theory of interactions ever proposed!* Yes, all of this stunning success of quantum electrodynamics requires the incorporation of renormalization, that at the outset admits our ignorance of the *sources* of the values of charge and of mass, but what an outcome! Well, the calculations have their get-your-hands-dirty aspects, but *renormalizable theories triumph.*

Theories for which we *cannot* put all of the vacuum polarization effects into the charge, mass (and perhaps some other observables) are nonrenormalizable and, at present, do not make any sense to us as *fundamental* theories. However, they are acceptable as *effective* theories, whose forms give us hints about the underlying theory for which they are good approximations in their realms of validity. Those realms are generally limited to energies below a characteristic value, above which the more general form of the fundamental theory must be used. There may be a series of more and more complicated effective theories (each containing the previous simpler one as an approximation) as we go up in energy, but the final fundamental theory that contains them all should be renormalizable in order to make sense to us.

4. WHAT ARE THE GAUGES IN THE GAUGE THEORIES?

The short answer is that there are none! The longer answer will send you on a historical quest to a theory created in the 1920s, which dealt with spacetime and measurement. Such a trip is too far off of our itinerary to pursue. So forget about gauges — a gauge symmetry is a *local* internal symmetry and the resulting gauge theory contains the requisite gauge particles and interactions!

<u>Gauge Theories</u>

(inspired by Beethoven's music to "Ode to Joy")

Sounds now sweep across the mountains, glorious tones that harmonize.
Sweet enchanting music flowing, with its contrapuntal ties.
Bringing us the thrilling message from the realm that 'bove us lies.
Fundamental interactions from these symmetries arise.

As we wander through the mountains that bring wonder to our eyes.
We're delighted and astounded as we start to realize
that these symmetries are "golden," far beyond our first surmise.
Fundamental interactions from these symmetries arise.

So the *beauty* that we cherish of the symmetries we prize
and *simplicity,* our idol, sought beyond all compromise,
come together in these visions, in a mathematic guise.
Fundamental interactions from these symmetries arise.

Fundamental interactions from these symmetries arise.

Now we see before us, shrouded in mist, two mountain peaks. The mist is so heavy that we cannot see beyond those peaks from here, nor can we tell whether they join together higher up. They are definitely in the realm of gauge theories, so our pause (Chapter 13) has provided us with useful information for our climb up to each peak. The air is already very thin at this altitude so we take deep breaths before our steep climb. Let us begin by climbing the peak of the strong interaction.

THE STRONG INTERACTION — QUANTUM CHROMODYNAMICS (QCD)

1. THE EXCLUSION PRINCIPLE MYSTERY

The puzzle

The original quark model described all hadrons, both baryons and mesons, as bound states of those fermions (the spin $\frac{1}{2}$ quarks). We expected, as the mathematics of quantum field theories tells us, that the quarks must obey the Pauli Exclusion Principle, i.e., two quarks cannot be in the same state. However, there are baryons (those hadrons composed of three quarks) that appeared to actually contain quarks in identical states. This puzzle had several possible solutions, one of which was to abandon the quark model, but of course nature would be the final arbiter.

Color and whimsy

The solution that nature has chosen for the exclusion-principle puzzle is that the quarks all have another quantum number. With *whimsy* working overtime, *they've chosen color for the name.* It is the *strong-interaction* analogy to electric charge (the particle property that causes particles to interact with photons), but there are *three different types* of this strong-charge called color. *Each quark flavor* can have any one of those *three colors*, so that there are three different *up* quarks, three different *down* quarks, three different *strange* quarks, etc. These colors are universally known as *red, green, and blue*, the primary colors. This choice of names has nothing to do with actual colors and everything to do with *whimsy* — our sometimes-misleading companion. After all, a different whimsical choice might have been made for this hidden property of matter, such as *fragrance*, for which the three types could be named *sweet, fresh and sour*. Of course, such a choice is not nearly as *colorful*. In fact, it *smells!*

The Scent (or Descent) of Color?

They've chosen color for the name.
They've chosen color for the name!
They've chosen color for the name
of strong-charge types,
of strong-charge types.

And what was wrong with "smell," pray tell?
Yes, what *is* wrong with "smell," *do* tell?
As "fragrance" it should sell, and well.
So blow the pipes!
So blow the pipes!

There's "sweet," there's "fresh," and "sour" too
to name those properties anew.
Let "smell" now make its grand debut.
The time is ripe!
The time is ripe!

We have not talked explicitly here about the antiquarks, but have implied that for each flavor of quark, there is a corresponding antiquark flavor, *e.g. anti-up* (not a verb; we ain't playing poker here!) also called u-bar (written \bar{u}) and *anti-down* (\bar{d}). Each *antiquark flavor* comes in three *anticolors*, so that there are three different *anti-up* quarks, etc. The corresponding anti-colors are referred to as *anti-red, anti-green and anti-blue*.* So, as we saw when we first entered the Land of Quarks, there are eighteen different quarks in nature: up, down, charm, strange, bottom and top, each in one of three possible colors.

ECHO FROM BELOW

And Don't Forget Their Antiparticles

We now *know*
three generations of quarks.
Even though
we cannot free them like corks.
The first one 's
in the stuff all around us.
The next ones
that we've found, don't confound us.

There are three
generations it is true,
and every
generation contains two.
That'd mean
that there are six kinds of quarks.
But eighteen
is the *colorful* answer, of course.

And what of the particles that *carry* the color interaction, those hidden gauge particles that provide the "glue" holding the hadrons together. Those *gluons* have a rather complicated color structure; technically they each carry one color and one

*It has been suggested that we replace anti-red by the color that results when we remove red from white light — a mixture of blue and green light called *cyan*. Similarly, we find the anticolors for green and blue to obtain altogether: *cyan* for \bar{r}, *magenta* for \bar{g}, and *yellow* for \bar{b}. Artists and stage-lighting experts are familiar with cyan and magenta, but I certainly am not. To my knowledge, these names are used in only one textbook and *nowhere* else in the physics literature.

anticolor. The effective color behavior of most of the gluons (six of the eight different types) is that they *change* the color of any quark that absorbs or emits them. In all reactions, the total amount of each color, including those of the gluons (counting an anticolor as -1), is always conserved.

The colorlessness of hadrons and friends

The reason that "color" was so hard to find is now well understood. The *hundreds* of different kinds of hadrons (both baryons and mesons) that have been found in numerous experiments are all *colorless*! Color appears to be a *hidden* property whose existence is extremely difficult to detect, even though it is responsible for the strongest force we have found in nature.

Let us consider mesons first. Although each meson is composed of one quark and one antiquark, the corresponding flavors of the two constituents need not match. For example, the (flavor) structure of the π^+ is $u\bar{d}$.* However, the *color* of the quark and the anticolor of the antiquark *do* correspond to one another, so that the meson is colorless.† This explicit cancellation of color occurs in all types of mesons.

<div align="center">

Headlines from:

THE MESON GAZETTE

CITIZENS CALMED BY NEWS THAT REDS ARE ALWAYS KEPT IN CHECK BY ANTI-REDS

INFLUENCE OF GREENS DILUTED BY ACCOMPANYING ANTIGREENS

TIPS FOR YOUR MENTAL HEALTH

When the *blues* strike you, remember that the *antiblues* will soon arrive.

</div>

Each *baryon* is composed of three quarks, without *any* antiquarks, so it is not apparent that those composite particles are colorless. However, as we shall soon explore, the color interaction is due to an SU(3)-gauge invariance called COLOR SU(3), usually written $SU_c(3)$. The colorlessness of baryons is an arcane result of the *intertwined-nature* of the SU(3) group, a property that we first encountered as we investigated the rotation group during our visit to the symmetry fair. The

*Its electric charge is $+1$, with $+2/3$ from the u and $+1/3$ from the \bar{d}.

†The wavefunction of each meson actually has equal parts of each color of quark accompanied by the corresponding anti-color of antiquark: $r\bar{r} + g\bar{g} + b\bar{b}$.

detailed mathematics of the group SU(3) shows that no $SU_c(3)$ operation can affect a state that consists of three quarks, each with a different color;* such a state is therefore a color singlet (it doesn't mix with any other states), which is a technical way of saying that it is *colorless*. Its colorlessness is analogous to the combination of red, green, and blue light that produces a white spot on your TV screen.

There appears to be another type of colorless creature in nature that possesses no quarks or antiquarks at all. It is a gluon-gluon bound state, which is called a *glueball* or, more formally, *gluonium*. Its existence has not been definitely confirmed, but the evidence keeps mounting.

Our present understanding of the $SU_c(3)$ theory contains a rationale for this "whiteness" of all observed particles. We will explore that explanation when we visit *color confinement*.

The Case of The Missing Colors (Sherlock Holmes & Dr. Watson)

Well Watson, I did not discern
their very strange predicament,
until you stated your concern
of possible imprisonment.

But Holmes, I had no solid clues,
and did not *know* they were confined.
Could not *imagine* nature's ruse,
and how they were so intertwined.

Exclusion violated *so*,
th' quark model I could not extol.
But I confess I did not know
strong interaction's *mighty* role.

But my good friend, I credit *you*
with pointing out what was quite plain.
The simple puzzle that you drew
let me their presence ascertain.

And they're all there in each hadron,
those colors "red" and "green" and "blue."
There deep within, *strong* goings-on
is so well hidden from our view.

And gluons never show up bare,
their colors in confinement's spell.
And even when no quarks are there,
gluonium does hide them well.

'Twas your idea, I must proclaim.
From just that small bit that leaked out
(*strong interaction* is its name)
you nailed it down Holmes, without doubt!

*The wavefunction of these colorless states has six terms, with the colors arranged differently in each term: rbg - rgb + bgr - brg + grb - gbr, in which the first entry in each term refers to the color of the first quark, the second entry in each term refers to the color of the second quark, etc.

The existence of the *three* colors shows up dramatically in experiments involving *intermediate states with real quarks*, since there are three times as many quarks as we first thought (eighteen instead of six). The rates for interactions involving those intermediate states are consequently *three* times as high as they would be for colorless quarks. Furthermore, when there are *virtual quarks* in the reaction calculations, the rates for interactions are *nine* times as high as they would be for colorless quarks. (The number *nine* results from the fact that virtual particles contribute to the probability *amplitudes*, whose squares give the probabilities, and three-squared is nine.) These large numbers (*three* and *nine*) have been confirmed by numerous experiments. They are *strong* (double entendre) confirmations of the "threeness" we call color.

<div align="center">Why Three? (Hercule Poirot & Capt. Hastings)</div>

I must tell you, Hastings, I'm very surprised
that Sherlock, he was not so curious too,
as I am about nature's scheme he surmised.
Yes why, mon ami, are there "red" "green" and "blue."

Yes Hastings, there are as you very well see,
without any reason that *Sherlock* did tell,
of such *strong* bold "colors" precise-ely three.
And that is disturbing the little grey cells!

> And yet, those experiments show without doubt
> with three times or nine times as much as would be.
> Such numbers are surely, Poirot, a great *shout*
> that *three* is the number, as you plainly see.

Yes three *is* the number, I do quite agree.
And rest assured Hastings that Poirot will show
why that is the number the colors *must* be.
I'll find that great principle from which they flow.

2. COLOR SU(3) — THE GAUGE THEORY OF QUANTUM CHROMODYNAMICS

The general structure

Now we are within the Standard Model and are climbing towards the top of the concept that explains the strong interaction, the existence of gluons and their properties, and the confinement of quarks and gluons. We have had a glimpse of the power and beauty of gauge theories, and will now look at the gauge theory that accomplishes all of this, "just by *being* there." That theory is called COLOR SU(3); it is the gauge theory of quantum-chromodynamics (QCD), where the prefix *chromo* means *color* (remember that *they've chosen color for the name*). The gluons are the requisite gauge particles and their interactions with the quarks are completely prescribed by the mathematics of the group.

In quantum electrodynamics (QED), the gauge particle is the photon. The photon has a basic interaction with electrically charged particles, but the photon itself has no charge. This chargelessness of the photon is a result of the simplicity of the gauge group of QED. (It is the group U(1) that has only one generator.) However, SU(3) has eight generators and they are intertwined. This complex intertwining causes the corresponding eight types of gluons to actually carry the strong charge, called color, in the intricate manner mentioned above. The gluons come in one color (of three) and one anticolor (of three), so we might expect nine different color-anticolor combinations, but the intertwined structure of this *simple* group (no "×"s in *its* name) gives us only *eight gluons* — the same number as the number of generators of SU(3).*

Color confinement

In order to appreciate the extraordinary behavior of the strong interactions between quarks (antiquarks) and gluons, it is useful to first look at the interaction between electrons (positrons) and photons. As we have seen, the gauge group that describes quantum electrodynamics has only one generator (and only *one* type of gauge particle — the photon) with the consequence that photons carry no charge. It is important to keep the chargelessness of the photons in mind as we explore here.

All charged particles have associated electromagnetic fields. These fields are actually quantum fields that are composed of virtual photons; however, here we are interested in their role as energy-storage "facilities." They are in this sense analogous to a spring, when a mass is attached to it, that stores (potential) energy. The configurations of those fields tell us about the nature of the forces exerted between particles.

It is simplest to consider the electric field that is present in a system consisting of one electron and one positron. That field extends over *all* regions of space, even reaching out towards infinity in all directions. An electric field is often depicted by representative lines, each of which must start on a positive charge and end on a negative charge, or extend to infinity. The strength of the field is apparent from the *density* of the field lines — where they are bunched closer together the field strength is higher than where they are more separated. In Fig. 14.2, we see a two dimensional sketch of the representative lines for a system composed of a positron (charged $+e$) and an electron (charged $-e$).

The electromagnetic field possesses stored energy, much more where the field lines are dense than where the field lines are more spread. As we pull the electron and positron further apart, the force we exert supplies more energy to the electromagnetic field. However, the mathematics shows that the amount of energy being "fed" into the field drops steadily toward zero as the particles are moved further

*Actually, the mathematics of gauge theories shows that the number of types of gauge particles *must* equal the number of generators in the symmetry group.

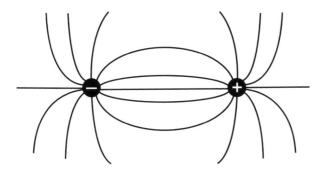

Fig. 14.2. A two-dimensional sketch of the electric field set up by a charge pair (electron and positron) with the density of the representative lines indicating the relative strengths. Notice that the lines are more widely spaced the further from the charged pair we look. They extend to all regions of space with the strength of the field falling rapidly (the density of representative lines dropping) as we go further and further from the particles.

and further apart. There is no difficulty moving them so far apart that the effective force between them is negligible compared to the effects from other particles in their respective neighborhoods. They then behave as "free" particles.

The corresponding discussion of the analogous situation for a quark and an antiquark (that form a meson) and their associated color fields is completely different. True, the gluons interact with the color property of quarks and antiquarks, just as the photon does with electric charge, but because the gluons themselves carry color-anticolor properties, they also interact with one another. Due to their color property, virtual gluon-gluon pairs are produced (as part of the vacuum polarization effect) in most of the *empty* space around the quark and antiquark, running all the way out to infinity. These virtual gluon pairs essentially *confine* the color-field lines that run from quark to antiquark to an extremely *narrow* region of space. This results in a very tight tube of force, called a *flux tube*, between the quark and antiquark, as shown in Fig. 14.3.

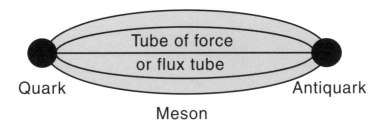

Fig. 14.3. A compact tube of force, called a *flux tube*, made of the virtual gluon lines running between the quark and antiquark of a meson. The "empty" space around the meson, and extending all the way to infinity, is inundated by the random production and destruction of virtual pairs of gluons (as well as quarks).

Song of the Vacuum Virtual-Gluon Pairs

We squash them together, together in bands.
Yes, squeeze them repel them, repel them from space.
We can't have *those* flux lines invading our lands.
Keep them in the flux tube, in that narrow place.

The strong intertwining that we have achieved
dictates that from flux tubes they may never veer.
This edict from nature we pairs have received,
so we won't allow them to trespass out here.

The amount of energy contained in the flux tube depends on its size and compactness. If we try to pull or knock the quark and antiquark further apart, we must supply more and more energy to "feed" the *lengthening* flux tube — its voraciousness is a consequence of its unchanging compactness. If we wish to get the quark and the antiquark so far apart that they are no longer bound to one another (a distance that is approaching infinity), we would need to supply essentially infinite energy. In contrast to electromagnetic forces, the effective *color force* due to the (color) flux tube between the quark and the antiquark *never decreases* no matter how far apart they are. In actual fact though, even when the quark and antiquark are still fairly close together there is so much energy in the flux tube that a new quark-antiquark pair can be produced. As the particles of the original pair move further and further apart, with the concomitant rapid increase in the energy stored in the field, the production of the *new* pair becomes more and more likely. As a consequence, a new quark pair is produced and the original meson becomes two mesons, as shown in Fig. 14.4.

Some of the stored energy, which had been supplied by the external force pulling the original pair apart (usually due to a particle fired at the original meson), has been converted into the mass energy of the *new* pair of quarks.

Thus, the inability to knock a quark or antiquark out of a meson results (indirectly) from the gluons' color content: The virtual pairs of gluons, constantly produced and destroyed in empty space as a part of the color-vacuum-polarization

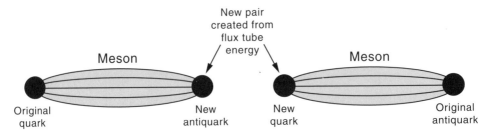

Fig. 14.4. Two mesons that result from an attempt to separate the quark and antiquark of an initial meson. The particles of the new pair, formed from energy stored in the flux tube, combine with those of the original pair to form the new mesons.

effect, *squeeze* down the room available for the gluon field lines extending from the quark to the antiquark. In a similar manner, *all* hadronic flux tubes (even in baryons and glueballs) are very compact, and quark-antiquark pairs and/or gluon pairs are produced from the energy supplied by an external agent. Consequently we can never knock the hidden color of hadrons out into the open. Furthermore, *no state of net color can exist* because some of the field lines emanating from the colored constituents would fill the otherwise "empty" regions of space all the way to infinity. It would take an infinite amount of energy for such field lines to displace, in (almost) all of space, the virtual gluon-gluon pairs that are there as part of the vacuum polarization effect. Only a color-neutral combination of particles can have field lines that form a compact region (not extending to infinity) of flux tubes, so only color neutral combinations of quarks and gluons (i.e., hadrons) or gluon-gluon bound states (i.e., glueballs) can exist.

ECHO FROM BELOW

Gluons

(Inspired by the folk song that starts "Freight train, freight train...")

Gluons! Gluons, holding so fast.
Gluons! Gluons, there is no blast
that can free the quarks you hold,
as all color you enfold.

Gluons! Gluons, holding so tight.
Gluons! Gluons, it is your plight
to be stuck inside as well,
in the same strong color spell.

The strong interaction between hadrons

The strong interaction between hadrons manifests itself all around us as the nuclear force holding all the nuclei of atoms together. That nuclear force is the strongest force of the four that we encounter in nature. We expect that that force is caused by the color interactions, but it is not immediately apparent how. After all, the color force does not decrease at all as quark and/or antiquark are moved further and further apart, whereas the strong interaction has a very short range — the nuclear force is nonexistent when particles are much further apart than the sizes of nuclei.*

Let us consider two hadrons (e.g., a proton and a neutron) that are near each other. We have just seen that all hadrons are color *neutral*, and this might lead you to think that there should be *no* strong force between them. You would be essentially correct if you came to that conclusion, but the situation is more involved when the hadrons themselves are roughly as close as the separations between their respective colored constituents (quarks, antiquarks, and/or gluons). In such a situation, a

*about one ten-trillionth (10^{-13}) of a centimeter. (Remember that a centimeter is about 39% of an inch.)

virtual gluon starting from a constituent of one hadron might stray and end on a constituent of the *other* hadron. The closer the hadrons are to one another, the more likely this is to occur. These occurrences will cause a distortion in each hadron's color distribution, and a concomitant force between them. Thus, this force is merely a *residual* effect of the "cloaked" color forces. This residual force is what was called the nuclear force for the first half of the 20$^{\text{th}}$ century and is still referred to as the strong interaction. We can see why this residual force between the hadrons is very short range *despite* the masslessness of the gluons and the "never-decreases" effect of compactness of the flux tubes (discussed above). It is because the *color-neutral* hadrons must be nearly as close as the separations of their constituents to feel it. And in fact, the *smallness* of the hadrons (the closeness of their constituents) has resulted from that very compactness of the flux tubes.

You might expect that each of the two mesons produced in Fig. 14.4 would have some of the virtual gluon lines of *their* flux tubes straying to the quark or antiquark of the other, with a consequent interaction between them. This may very well happen but once the new mesons have sped away from one another nothing more can occur. Now if we *intentionally* bring two hadrons very close together (e.g. by "firing" one at the other), we certainly might find hadronic arrangements of quarks and antiquarks emerging that are *different* from the initial hadrons. Often the virtual gluons involved have produced *new* quark-antiquark pairs. This explains our preference for the word "interaction" to describe the strong goings-on. No matter what the process though, only *colorless* combinations finally emerge.

In high-energy collisions, hadron beams are generally fired directly at one another. Occasionally, two hadrons, one from each beam, collide. From the analysis of the results of such experiments, it appears as if *one* colorful constituent (a quark, antiquark, or gluon) from one colliding particle scatters strongly from *one* colorful constituent of the other colliding particle, with their other constituents behaving as "spectators" of that *hard scattering*. The strongly scattered constituent particles generally have received large "kicks" out of the beam directions. However, *no* quarks, antiquarks or gluons can emerge with their colors uncloaked. A multitude of processes, involving virtual particles, must follow to produce the colorless hadrons (with the colored particles "locked" inside) that subsequently emerge from the interaction. These processes are collectively referred to as *hadronization*. Although their detailed courses have not been deciphered, we know from observation that the two strongly scattered constituents finally do emerge (usually moving out sidewise to the beam direction) each locked in one of the particles of a *jet* of hadrons. Each corresponding jet has a total sidewise momentum that is very nearly equal to the large "kick" that the scattered constituent received from the hard scattering. These jets are sometimes cited as experimental observations of quarks and gluons; they *are* the best we can do because of color confinement. In fact, no collision can result in the emergence of a free quark or gluon because of confinement, so that *hadronization* must occur in any *hard* collision involving a hadron.

Those magic hands that nature wields
they're fast and never seen.
Struck quarks from view they quickly shield
and all the colors screen.

The same for gluons, just as fast
they do it, oh so quick,
right after any *strong* hard blast,
th' hadronization trick.

That hadronization trick is "cool",
its intricacies so slick.
But we now know that it's the rule,
though we can't *track* the trick.

3. A NEW *PHASE* OF MATTER

In our discussion of the quantum properties of nature, we used the word "phase" to refer to a location on a waveform as well as to describe the relationship of interfering waves (e.g., in phase, out of phase). This word "phase" is also used in describing the state of a substance (like water, H_2O), where it has a *completely different meaning*. — Should we put a check (mark) on his check, or would you take a rake to the rake? — Consider water and the different forms it can take: at low temperatures it is a solid (ice), at higher temperatures it becomes a liquid (which is usually meant when we use the word "water"), and when it boils away it has changed into a gas (water vapor or steam).

The H_2O molecules of water vapor are pretty much independent of one another, rushing here and there quickly and occasionally colliding. As we cool water vapor below the boiling point (212° Fahrenheit or 100° Centigrade) a profound transition takes place. The molecules move more slowly so that their weak pulls on one another are effective enough to cause bunching into drops. The drops fall and gather into puddles that we recognize as liquid — the dew wets the grass on a cool night in fall. An "Alberta Clipper" (brrr) moving into the Midwest can quickly cause those molecules to align into a fancy, rigid pattern, so that they no longer can flow; this transition has taken the water from liquid to solid. Each of the different states of water (solid, liquid, gas) is referred to as a different *phase* and the change from one phase to another is called a *phase transition*. (In fact, further lowering the temperature of ice will take the ice through several more phase transitions to different patterns of the molecules referred to as 'different ice phases.') So, we use the word *phase* to refer to the markedly different behaviors (as solid, liquid, gas, etc.) of the *same* substance when it is at different temperatures and/or pressures. We understand the phases to correspond to different arrangements of its constituents (molecules).

The hadronic matter we see all around us, consisting of those bound bundles of quarks and/or gluons that we call mesons, baryons, or glueballs, is sometimes said to be in the *hadronic phase*. The constituents are bound together by the very strong color forces, *confining all color* so that it is completely hidden. Many people wonder whether this binding might be broken at very high temperatures. We borrow the language used for macroscopic substances to ask: If the universe were raised to a very high temperature and/or if matter were squeezed together to produce very high densities, would there be a *phase transition* of matter from the hadronic phase to an unconfined-color phase? In other words, would colored particles (quarks and gluons) be pretty much independent of one another, rushing here and there like the molecules of a gas? Ah, but we believe that the universe *has* gone through such a change of temperature and density, although in the opposite order. All evidence indicates that it started as an exploding point (presumably at infinite temperature with all the energy squashed into that point); that explosion is called the *Big Bang*. In less than the blink of an eye the energy had converted into extremely hot matter that we believe was in the form of quarks and gluons (as well as leptons). The quarks and gluons were not bound together in hadronic bags, but instead were freely moving; matter was in an unconfined-color phase, often referred to as a *quark-gluon plasma*. The universe rapidly expanded and cooled until a phase transition occurred leaving strongly interacting matter in the hadronic phase that we find it in today.

In order to test the validity of this scenario of a phase transition from quark-gluon plasma to the hadronic phase, scientists are planning to "recreate" the presumed conditions of the very early universe. They hope that in a very small region for a very short time they can bring ordinary matter to very high temperatures and densities. Although we can only expect such an attempt to succeed for an extremely short time, the subsequent behavior of that matter might very well indicate that a quark-gluon plasma had been produced. A huge collider (that accelerates two beams of particles to high speeds and then smashes them together) has been built in a tunnel on Long Island (NY), in order to smash two big nuclei into one another (such as gold on gold). It is hoped that so much matter (two gold nuclei together contain 394 nucleons), smashed into a very small space and containing so much (kinetic) energy due to their high speeds, will approximate the (presumed) conditions of the early universe well enough to produce a quark-gluon plasma for an instant in that very tiny region. If all goes as planned, this may soon be accomplished.*

*The large particle accelerator in Geneva (CERN), of much lower energy, has very recently reported data indicating that they *may* have created a quark-gluon plasma for an extremely short instant.

4. MORE *EXOTICA*

Hybrid and Exotic States

The existences of a few strongly interacting particles (hadrons) that do not fit the standard structures of the Standard Model, quark-antiquark (for mesons) and 3 quarks (for baryons), were found in the late 1990s. Some of these appear to be hybrid states that contain coherent superpositions of standard and non-standard structures. Others have properties* that show that they cannot contain standard structures at all; they are referred to as *exotic states*.

Breaking News - April 10, 2002

NASA's Chandra X-ray Observatory looks for evidence of the existence of neutron stars, which are expected to result from the collapse of heavy stars after their nuclear fusion reactions have ceased. By combining its data with those of the Hubble Space Telescope, researchers have found evidence of an object that is too small to be a neutron star, but could be construed as our first observation of *stable* matter containing *strange* quarks. However, this is very preliminary and further careful analysis must be done to rule out less exotic interpretations such as a normal neutron star with a hot spot. They also have evidence that another star is too cool (in the temperature sense of the word, of course) to be a standard neutron star; first guesses include the possibility that it is made of a form of matter, neither seen nor contemplated for stars so far. So, STAY TUNED!

*angular momentum, along with behavior under reflection and charge conjugation (to be discussed in Chap. 15).

QCD Melody

We sing the song of QCD,
the *strong* unbroken gauge theory
whose structure 's local S-U-3,
a song of secret revelry.

From quark to quark, its gluons fly,
their colors' needs to gratify.
Though locked in tubes, their dance we spy,
but we can't see them eye-to-eye.

This property that comes in three
's confined securely, never free.
Obscured is so much pageantry;
yes, hidden 's all the poetry.

So sing the song of QCD,
sequestered local symmetry,
a soaring joyous rhapsody
of nature's wondrous artistry.

We have just become acquainted with quantum chromodynamics (QCD), which is the underlying gauge theory of the strong interactions, and have previously talked about the gauge-theory structure of quantum electrodynamics (QED) as well. Now we will attack the formidable weak interactions, which, after many baffling twists and turns, will lead us to the electroweak theory. That theory will show that the electromagnetic and weak interactions are intimately related to one another.

THE WEAK INTERACTION — THE FOOT-HILLS

1. BETA DECAYS

Beta (β) decay and neutrinos

We have already visited the phenomena called beta decay, but let us pause at the beginning of our climb to recall that experience and reflect upon it. Beta decay is the name for radioactivity in which an electron or positron comes flying out of an atomic nucleus, with the concomitant change of that nucleus into one with a higher or lower charge, respectively. (The total charge of the resulting nucleus plus outgoing charged lepton is the same as the charge of the *parent* nucleus.) Our firm belief in the conservation of energy, momentum, and angular momentum led us to understand that in each such decay an electrically neutral, *apparently* massless particle called the *neutrino* also emerges from the nucleus. We believed this during the two decades that followed the suggestion even though the neutrino had not been experimentally detected. We understood that it was undetected because of the properties of the neutrino itself; it is uncharged and only interacts with other particles via the weak interaction. It would require tons of tons of \cdots detector material to have a reasonable chance of detecting the single neutrino emerging. Our faith (oops) in those conservation laws, which result from the space-time symmetries we visited earlier, has been vindicated by experiments using tons of detector material at a nuclear reactor (where the flux of antineutrinos is very large) — the expected rate of antineutrino detection *was measured*. (Since that time, many neutrino detectors have come on line, but the intriguing phenomena they have discovered are another story, which we will hear after we have mastered the Standard Model.) BRAVO for the symmetries and conservation laws; BRAVO for our faith (there's that word again) in them; and BRAVO for the creator of the neutrino conjecture (in the 1930s) and the researchers who actually *found* them (in the 1950s).

The simplest β-decay that occurs is the decay of the neutron; it decays into a proton, an electron and an electron-type antineutrino ($n \rightarrow p + e^- + \bar{\nu}_e$). Neutrons

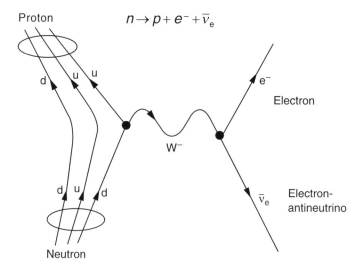

Fig. 15.1. The fundamental particles' *effective* behavior in neutron decay. The neutron is a bound state of two d quarks and a u quark, and is represented by the oval in the lower left corner of the diagram. The proton is a bound state of two u quarks and a d quark, and is represented by the oval in the upper left corner of the diagram. The right-most d quark in the diagram is undergoing a weak decay, becoming a u quark as it emits a virtual W$^-$. The W$^-$ produces an electron and an (electron-type) antineutrino.

are usually well behaved in a nucleus, but just get a bunch of them alone (free). They will desert us every time, with about half of them decaying every ten minutes. After a while, all we have are protons and electrons to show for our trouble (the antineutrinos being too elusive for us). Our interpretation of the fundamental process occurring during β-decay comes from our understanding of the weak interactions. The neutron that decays is really a "bag" containing quarks (two d quarks and one u). *Effectively*, it is one of the d quarks that changes flavor becoming a u quark and emitting a W$^-$ in the process. That (virtual) W$^-$ produces an electron and an antineutrino, as shown in Fig. 15.1, commonly called a "quark-flow diagram."

This diagram symbolically shows the quarks and leptons involved and the virtual W required to produce the decay. The baryon number (equal to 1/3 the number of quarks) is manifestly unchanged. The lepton number is unchanged as well — the electron has lepton number +1 and the antineutrino has lepton number −1, so that the total final lepton number is zero just as it was for the initial state (the neutron).

This kind of decay can occur in nuclei as well; effectively, a nucleon in a nucleus decays and the resulting leptons come flying out. In some, the decaying nucleon is a neutron, with an electron and antineutrino emitted as above, and in others a proton (effectively) decays with the emission of a positron and neutrino. A well-known example of a radioactive nucleus that beta decays is carbon 14, used for dating ancient objects. All carbon nuclei have 6 protons; the corresponding 6 electrons

that surround the nucleus determine that all such atoms behave chemically alike — they are all carbon. The "14" tells us that there are 14 nucleons in the nucleus; since 6 of them are protons, the other 8 are neutrons. This is not a stable form (technically, *isotope*) of carbon; the stable isotopes of carbon are carbon 12 (6 protons plus 6 neutrons) and carbon 13 (6 protons plus 7 neutrons). Carbon 14 beta decays to nitrogen 14 (7 protons plus 7 neutrons) as the nucleus emits an electron and an anti-neutrino: $^{14}C \rightarrow {}^{14}N + e^- + \bar{v}_e$. Effectively, a neutron within the nucleus has decayed to a proton. On the other hand, another isotope of carbon, carbon 11, emits a positron (and neutrino) to become boron 11 (5 protons plus 6 neutrons), as if a proton in the nucleus has decayed into a neutron (and a virtual W^+): $^{11}C \rightarrow {}^{11}B + e^+ + v_e$. A *free* proton cannot change into a neutron because the neutron is more massive. Actually, a free proton does not decay at all, as far as we know; since it is the lightest baryon, it would have to decay into non-baryonic matter, thus violating conservation of baryon number. We will learn more about the question of proton decay when we visit *grand unified theories*.

Other hadrons undergo β-decay as well, with one of the quarks or antiquarks decaying into a quark or antiquark of a different flavor, respectively, and emitting a virtual W in the process (as in Fig. 15.1). That W becomes an outgoing pair of leptons (a lepton and an anti-lepton) — a positron or electron with a neutrino or antineutrino, respectively.

<div align="center">

Beta Decay

Beta decay, yes its beta decay —
zooming electron or positron ray.
Stealth-like, neutrinos fly out th' other way.
Beta decay that ol' beta decay.

Not at all subtle is beta decay,
W gaugons' bold "virtual" trick.
Flavor 's changed, charge has been carried away,
every time that Geiger counter goes "click."

Beta decay, yes its beta decay —
zooming electron or positron ray.
Stealth-like, neutrinos fly out th' other way.
Beta decay that ol' beta decay.

Unstable nuclei, restless no doubt.
Radioactive, they each will decay
just as new leptons come flying right out.
Even a neutron will vanish this way.

Beta decay, yes its beta decay —
zooming electron or positron ray.
Stealth-like, neutrinos fly out th' other way.
Beta decay that ol' beta decay.

</div>

$$\mu \rightarrow \nu_\mu + e^- + \overline{\nu}_e$$

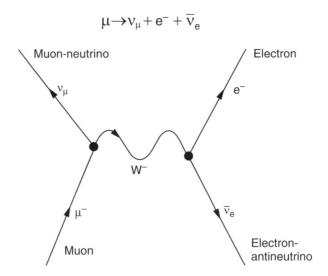

Fig. 15.2. Muon decay, showing the decay of a μ^-. A similar diagram can be drawn for the decay of a μ^+ (the anti-particle of the μ^-): the outgoing particles would be the anti-particles of those shown, with a virtual W^+ carrying the charge to the outgoing positron.

2. THE OTHER WEAK INTERACTIONS

A (purely) leptonic weak interaction

The simplest leptonic weak interaction is muon (μ) decay. (Remember that the muon is the second-generation charged lepton.) The muon is unstable, with a probability of $\frac{1}{2}$ to decay in about 2 millionths of a second. In this decay a μ^- changes into a mu-neutrino, with the emission of a virtual W^-. That W then produces an electron along with an electron-antineutrino. See Fig. 15.2.

(Completely) Non-leptonic weak interactions

There are weak interactions in which both ends of the virtual W, that "transmits" the weak interaction, are connected to quark lines. A typical example is the decay of a Σ^- (sigma-minus, which is a bound state of an s and two d quarks) into a neutron and a π^-. The corresponding quark flow diagram is shown in Fig. 15.3.

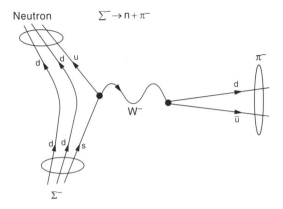

Fig. 15.3. The fundamental particles' behavior in Σ^- (sigma minus) decay. The Σ^- is a bound state of an s, and two d quarks; it is represented by the oval in the lower left corner of the diagram. The π^- is a bound state of a d and \bar{u}. The s quark undergoes a weak decay, becoming a u quark as it emits a virtual W^-; consequently, the strangeness changes in this decay. The W^- produces a d quark and an anti-up antiquark (\bar{u}), bound together as a π^-.

The Double-Trouble Ws

They're shrewd mischief makers though never overt,
those W brothers, those two.
They flip the quark flavors and quickly convert
to leptons while leaving no clue.

We know they did strike, yes we know without doubt
just *how* they can manage so well
to cover their tracks as they scurry about,
the weak beta decays to compel.

Or even more subtle decays they can cause,
converting to quark pairs instead.
No leptons appear, yet that won't give us pause.
We know how to tell where they've tread!

In case you are skeptical let me assure.
We know where they "live," where they're free.
We've made some *appear* with high-energy lure
and studied detailed properties.

Note that we have not seen any Z^0 interactions so far on our journey, even though you have already been told that the gauge particles of the weak interactions are the Ws *and* the Z. The Z does not mediate any flavor changes and its interaction is so well hidden that the evidence of its existence was not seen until monumental experiments were designed and performed in the 1970s. Since then many millions of Zs have been produced and studied in detail. It was the electroweak gauge theory (created in the 1960s) that predicted the existence of the Z and stimulated those experimental endeavors.

3. HANDEDNESS, TIME REVERSAL AND ANTIMATTER

Left-handedness is not sinister (from the Latin for "left")

Theorem: Left-handedness is not sinister. *Proof*: I am left-handed but no one has ever accused me of being *sinister*. In fact, I am definitely not sinister! *Q.E.D.* (Thus it is proven!) It's just that we've had millennia of bad press — that old *right-hand press*! A more important fact is that nature itself is left-handed. (Of course, this is the "spin" that this lefty is putting on the facts that follow.)

The electrons, positrons, neutrinos and antineutrinos emerging in β-decays are all spin $\frac{1}{2}$ particles. This means that each can have spin up or spin down along its direction of motion — its momentum direction. The spin angular momentum of a spin up particle is analogous that of the rotation of your right hand as you insert a screw, with the screw driver (and your thumb) pointing in the particle's (and the screw's) direction of motion. Therefore, spin-up particles are also referred to as *right-handed*. Likewise, particles with spin down along their momentum directions are called *left-handed particles*. Every β-decay ever measured has had either a *left-handed neutrino* or a *right-handed antineutrino* emerging, but *never* have we found a right-handed neutrino or a left-handed antineutrino. In fact, in the Standard Model, the neutrinos are massless left-handed particles; *no right-handed neutrinos exist at all.*

Standard Model's Neutrino Song

Neutri-no, Neutri-no,
flying by so fast.
Guess I know, yes I know
why you last and last.

Neutri-no, Neutri-no,
flying at high speed.
Guess I know, yes I know
y'll go right through, indeed.

Neutri-no, Neutri-no,
massless (so you seem).
Fly, fly, fly;
zoom on by,
'n glorious *lefty* beam!

The high-speed electrons (leptons) emerging have shown a very strong tendency to be left-handed, whereas the high-speed positrons (anti-leptons) are overwhelmingly right-handed. (There is a technical detail about particles with mass that precludes *all* of the electrons from being left-handed and *all* of the positrons from being right-handed.) This is part of the evidence that the weak interaction acts only on *left-handed leptons* and *right-handed anti-leptons*. There is now overwhelming evidence that the Ws and Z, the propagators of the weak interaction, interact only with left-handed quarks and leptons. So, lefties are special after all!

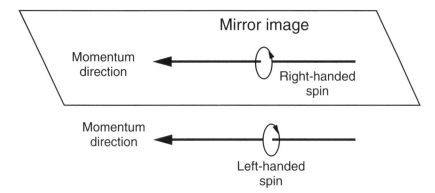

Fig. 15.4. A representation of the spin orientation of a particle with respect to its momentum direction. Since the spin of the (left-handed) particle is rotating *away* from the mirror's surface, so must the spin of its image (i.e., it is rotating into the mirror). Thus the mirror image looks like a right-handed (ordinary) screw when the *particle's* spin has a left-handed-screw orientation.

The definite handedness of the weak interactions tells us that if we were to see the reflections of weak-interaction experiments in a (perfect) mirror or the recorded data of such a reflection observation, we would know that the mirror was there. The reported data could not have been directly from an experiment. Very simply, in the mirror image all neutrinos and other leptons would be right-handed — all handedness properties of the weak interactions would be reversed. This shift of left-handed to right-handed may be familiar to you; just raise your left hand while looking into a mirror — your image has its right hand up! For the same effect on spinning particles, see Fig. 15.4.

We express the impossibility of the mirror-image data being the direct description of any experiment in our world by saying that *there is no reflection symmetry in the universe.* The technical name of the mathematical operation that changes the world into its mirror image* is *parity* and it is represented by capital P. Note that we can always take the data of an experiment and change them into mirror-image data, which is equivalent to applying the P operator; however, for weak interactions the changed data *cannot* correspond to the direct observations of *any* experiment.

The handedness we find in the weak interactions may not surprise you because we see handedness differences around us quite often, e.g., the preponderance of righties (people), growth patterns of definite handednesses in different species of animals (like snails, etc.) and plants, and the fact that proteins are made of left-handed amino acids wound in right-handed chains. However, it may surprise you to know that for centuries physicists believed that the basic *underlying* interactions and particles have no preferred handedness (we would say that they thought P was a symmetry of the world), and that the evidence to the contrary all around us was due

*Actually, the parity operator also contains a rotation of 180° about the reflection direction, so that it is equivalent to inverting everything through the origin of the coordinate system.

to arbitrary choices made (unknowingly) by the particular species here on Earth. The suggestion that P invariance was violated by the weak interactions was made in 1956 and verified by several experiments shortly thereafter. This discovery startled the physics community, which had unquestioningly believed in P invariance.

Matter-antimatter identity interchange (C)

Now let us suppose we examine the data of a weak-interaction experiment, but that the researchers fib and say that the particles involved were the antiparticles of the actual particles in the experiment. (They also clandestinely reverse all electric and magnetic fields in the experiment, as if they had been produced by the flow of positrons instead of electrons.) Could we tell that they were playing that trick on us? Well, of course! They would be calling the left-handed neutrinos "antineutrinos" but we *know* that all (weakly-interacting) antineutrinos are right-handed. Likewise, they would call the right-handed antineutrinos "neutrinos." The interactions with *leptons* are all *left-handed* and they would claim that those were antileptons, for which the interactions should be right-handed; likewise, they would have leptons experiencing the *right-handed* interactions that the antileptons should experience. It's as easy as π to tell!*

The letter C is used to represent this matter-antimatter identity interchange. So we can say that *the weak interaction, and hence the world, is not invariant to C.*

The concept of antiparticles began with the prediction and subsequent discovery of the positron, the antiparticle of the electron. Replacing one by the other amounts to reversing their charges (e becomes $-e$ and vice versa). The operation C was therefore given the name *charge conjugation*. Unfortunately, this name has stuck, even though for most particles there are other chargelike (additive) numbers that must be reversed as well, *e.g.* baryon number, lepton number, I_z, hypercharge, strangeness and charm. (The corresponding numbers for the highest generation of quarks would be "topness" and "bottomness", but these names have never been used since, by the time they were discovered we already knew how to keep track of the *quark* content of particles.)

CP or not CP, that is the question

Our arguments above against P and C invariance were based on the fact that the *particles* must be associated with *left-handedness* and their *antiparticles with right-handedness*. Now suppose those practical jokers, who have been giving us "doctored" data, changed all particles into their antiparticles and also replaced the data with "mirror-imaged" data. We would see left-handedness associated with the "particles" and right-handedness associated with their antiparticles — nothing wrong with that! You might expect that this combined operation (CP) would be

*Just kidding. There is an old expression for something that's a 'no-brainer': "It's as easy as pie!"

an invariance of the world, even though C and P separately were not. Everyone did! However, in 1964, experimenters found a very small but nonzero effect that violated CP, which they weren't even looking for — serendipity. So the answer is "not CP", but only "not" by a very small amount.

Time-reversal (T) — running the film backwards

Suppose that our mischievous tormentors made a film of an elementary particle interaction but ran it backwards when they showed it to us. This would correspond to applying time reversal (T) to the experiment.

If I would run my favorite *movie* backwards for you, you would immediately know what I had done — people don't walk backwards, beverages don't emerge from people's mouths to fill empty bottles and balls bouncing around on a pool table don't finally form a triangle at rest with only one rolling away from the apex of a triangle of pool balls. But could people walk backwards, could liquids emerge from our mouths to fill empty containers, could we find the right velocities for the pool balls (by studying the breaks in pool and suitably reversing all of the final velocities)? I guess so. Certainly those events would be much much more *unlikely* to occur in that (time) order than in the order that they usually do.*

It had long been believed that the *fundamental* interactions of nature look the same when "we run the film backwards." This implies that any elementary particle event, when run backwards, should look like another possible event. However, the "not CP" discovery in 1964 casts a dark shadow over the possibility that the weak interactions are invariant to time reversal:

A theorem to remember — CPT invariance

A beautiful, elegant mathematical argument shows that, if we were to perform all three operations C, P, and T on *any* acceptable[†] relativistic quantum field theory, it would look unchanged when we were done. In other words, if our wise-guy experimentalists were to film their events through a mirror, run the film backwards, and tell us that each particle in the reaction was the antiparticle of what actually was there, we would have no way of detecting this compound ruse. The formal theorem says that *all (local) relativistic quantum field theories are invariant to CPT*. (This was once called the "TCP" theorem, but when that fuel additive was widely advertised the name became CPT.)

An important consequence of this CPT theorem is that if T alone were an invariance of the basic interactions, then so would CP have to be as well. Just consider any situation where reversing the film is undetectable; then the subsequent charge

*In fact, just such a probability argument in the theory of the molecular structure of matter called *statistical mechanics* is the justification of the second law of thermodynamics.

[†]that does not have spread-out-in-space (called "non-local") basic interactions.

conjugation combined with the parity operation must leave the interactions finally unchanged, because the CPT theorem says that all together these three operations definitely do not change them. This has the consequence that the 1964 detection of a violation of CP invariance implies a violation of time-reversal invariance (T). In fact, people often refer to the CP-violating data as evidence for a violation of T, even though these are completely different operations. *As CP goes, so goes T!*

<div align="center">

CPT CP T

</div>

Narrator:	And from above does thund' resound
	proclaiming CPT to all.
	That combination nature crowned
	to reign unbroken — never fall.
Narrator:	And 'tis its reign that is the twine
Chorus whisper (p)	CPT CP T
Narrator:	so tightly binding T 'n CP;
Chorus louder (mp)	CPT CP T
Narrator:	their fates so linked by bold design,
Chorus louder (mf)	CPT CP T
Narrator:	completely different though they be.
Chorus loud (f)	CPT CP T

4. RANGE, STRENGTH AND MASS

Usually, the range of a force or interaction is smaller, the larger the mass of the particle transmitting it. In our most recent climb, we have seen that this rule of thumb does not hold for the strong interaction; the gluons are massless but the observed interactions (between our colorless friends the hadrons) are very short-ranged due to color confinement. There is no confinement of the leptons, nor of charge, so that the connection between mass and range does hold for weak and electromagnetic interactions: The weak interaction has a very short range and the masses of the Ws and Z are certainly enormous (about 86 and 97 times the mass of the proton, respectively); the electromagnetic interaction is technically infinite-range and the mass of the photon is indeed 0. Furthermore, the weakness of the weak interactions (compared to the electromagnetic) can be traced back to the heaviness of the W and Z particles, since at least one of them is exchanged between weakly interacting particles.

Although the connection between mass and range is correct here, those very *masses* of the Ws and Z form a major road block to our understanding. We have looked earlier at the general structure of gauge theories (before our exploration of quantum chromodynamics) and noted then that the gauge particles of a gauge symmetry are *massless*. We did talk about the renormalization of mass (and charge), which is the only way we know to make fundamental gauge theories meaningful; however, it cannot account for the non-zero masses of the gauge particles, the Ws and Z. The renormalizations do *not* break the gauge symmetry and that symmetry

guarantees that the gauge particles remain massless. So the short range of the weak interaction and the corresponding enormous masses of the Ws and Z deter us from understanding the weak interaction in terms of a fundamental gauge symmetry. We shall soon learn the hidden secret that reconciles this conflict.

Since the sun is going down, it is time for us to make camp and rest before tomorrow's steep and labyrinthine climb.

Stretching before us is the mass morass resulting from the short range of the weak interactions. It took a decade of thought and effort to find a mathematical trail leading around this swamp, and we will now follow the path of concepts that was developed as part of that trail. It will take us through a region inundated by mist, where we cannot be certain of the contours or vegetation. However, our footing will be steady enough so that we will finally emerge into the sunlight.

Chapter 16

THE ELECTROWEAK INTERACTION AND THE STANDARD MODEL

1. HIGGS MAGIC

The mass morass

The Ws and Z seem to fit our notions of the role of gauge particles in the transmission of the weak interactions. Furthermore, there are indications that the (very massive) Ws and Z are closely related to the gauge particle of the electromagnetic interaction, the (massless) photon. But any attempt to modify a gauge theory by slipping some mass in for these particles would lead to an unrenormalizable theory — renormalizable quantum field theories for massive spin 1 particles, like the Ws and Z, *do not even exist*. And an unavoidable requirement of all quantum field theories, if they are *fundamental* theories, is renormalizability. Unrenormalizable *fundamental* theories make no physical sense to us. At this point, as we follow the path of the creators of this theory, we are "knee-deep in a big muddy!"

<u>No gauge particles?</u>

There are no ga-auge particles in here.
There are no ga-auge particles in here.
'Tis easy to show that we've all got mass;
we're sorry you're stuck in that great mass morass.
These facts decry your theories, 'twould appear.

There are no ga-auge particles in here.
There are no ga-auge particles in here.
We're proud to proclaim that our masses are huge;
you'll have to uncover the sly subterfuge,
to find the *real* gauge particles in here.

There are no ga-auge particles in here.
There are no ga-auge particles in here.
But if you are willing to travel *around*;
the subtly hid secret will surely be found,
the fate of *those* gauge particles in here.

Avoiding the morass — *Higgs magic*

Now a mysterious magician covered in shadows appears, beckoning us to follow him into the mist around the swamp. Let's see what he has in mind. He shows us two identical-looking boxes, one marked W^+ and the other marked W^-, floating in the air. He now opens the front and back doors of each so that we can see right through them, as magicians usually do. They are both definitely empty! He closes them and waves his shadowy arms in front of them while doing a strange jig accompanied by an eerie chant. After stopping he opens the boxes — they are full of sand! (applause)

Now the shadowy apparition rolls up his sleeves and pulls two very clean handkerchiefs out of his pocket, one made of cotton the other of linen. They both have the word *neutral* embroidered across in large black letters. He shows us that aside from those two handkerchiefs his hands are empty. With his left hand, he slowly threads the cotton handkerchief into the top of his closed right hand and then does the same with the linen one. They are both in there. After performing that strange jig and chant again, he begins to slowly pull a handkerchief from the bottom of his right hand. Behold; it is largely linen but *some cotton* is interwoven. It is heavy, covered with moist clinging sand! He pulls a second handkerchief from his hand that is mostly cotton with some linen woven in. It is absolutely clean! Each of these has the same word *neutral* embroidered across its entire area. He then opens his right hand to show that it is now completely empty and clean — no trace of sand in it. (standing ovation)

The W^+ and the W^- definitely emerge from the fog with mass! The massive Z^0 (the sandy one) and the massless photon (the clean one) also appear before us! The masses of the Ws and the Z have been produced by the Higgs magician. We are reaching out (experimentally) to grasp this elusive vision, but so far he has evaded our touch. Enveloped by the mists, we wonder: Does the Higgs magician really exist? If so, how does he do his magic? Why has nature created this magician and his magical performance? There is a mathematical conjecture about *how* this apparent magic might occur. The answer to the *why* question is much more elusive.

The Higgs magician has produced a weird breaking of the gauge theory that does not affect the symmetry of the *interaction*, but changes some of the massless gauge bosons *into massive* physical particles (Ws and Z). Actually, this symmetry breaking appears in *all* physical states but not in the interaction! This Higgs magic is referred to as *spontaneous symmetry breaking*. We can experience spontaneous symmetry breaking in a simpler form as we eat the perfect bagel. Before I take the first bite, the bagel has a rotational symmetry: Keeping it flat on my kitchen table, I can rotate it about the center of the hole by any amount and it still looks unmoved. As I take the first bite, although choosing where to bite *arbitrarily*, the rotational symmetry of the bagel is *destroyed*. The location of the bite would tell the tale of any rotation! If every day, for thousands of days, you put an identical perfect bagel at that place in front of me each morning, my first bites would be randomly

distributed around the circumferences, because the "law" of my bagel-nature is rotationally symmetric. Yet, if I have only *one* bite "to give for my country," that bite breaks the symmetry of the bagel-world. The bitten bagel is analogous to the physical states that do not have the symmetry in question, even though my bagel "laws" that are analogous to the interactions still possess the symmetry. If the Higgs does indeed exist, we ask: How and why does the Higgs take its bite out of the physical states? Has it taken the *same* bite in every part of the universe? The cause of this spontaneous symmetry-breaking is unknown!

The gauge theory relevant to the weak interactions is based on the SU(2) of isospin pasted together with a U(1) of hypercharge (both generalized to include leptons). The left-handed quarks and leptons and their right-handed antiparticles interact with the isospin gauge particles; the right-handed quarks and leptons and their left-handed antiparticles do not.* Consequently, the electroweak gauge theory is referred to as $SU_L(2) \times U(1)$, the sub-L (sometimes omitted) reminding us of the handedness properties.

The mathematical trail we are following has an extra particle field called *the Higgs field* (Tah-tah!) — our nebulous magician. And that Higgs field is present everywhere. Even in the hypothetical vacuum state, the Higgs field may be found permeating all of physical space. This is not like the virtual pairs that we renormalize away. It is *physically* present in such a way that the vacuum state and all other conceivable (physical) states of the world *break* the SU(2) × U(1) symmetry. — *The interactions themselves obey the symmetry* but the (*physical*) *states* emerging from the Higgs magic *do not*. Because the interactions still obey the symmetry, the quantum field theory is renormalizable!

The gauge particles of the *unbroken* SU(2) × U(1) electroweak theory do not have mass, indeed they cannot have mass! However, we have "seen" that they interact with the magical, nebulous Higgs field in such a way as to drastically modify their behavior. The Higgs field causes the charged gauge particles (W^+ and W^-) of the SU(2) to *behave* as if they had mass. (Remember the boxes of sand.) They have an *effective* mass, which is responsible for the short range of the interactions they transmit as *virtual* particles — the connection between range and mass that we have previously talked about. This effective mass of the Ws is *directly* observed when they are produced as *free* particles. So, when we say that the Ws are massive gauge particles, we secretly nod our heads to one another to indicate that we are aware of the elaborate symmetry breaking (Higgs mechanism) that produces their massive behavior.

There are two electrically neutral gauge particles in the SU(2) × U(1) theory: a W^0 (the handkerchief of linen) in the isospin SU(2) and a neutral gauge particle (the handkerchief of cotton), the gauge particle of the hypercharge U(1). The

*Technically: The left-handed quarks and leptons (and their right-handed antiparticles) form the isodoublets we talked about in Chapter 12; the opposite handed particles are each a separate isosinglet.

Higgs field has done something even more astounding to them than it has to the charged Ws. It *intertwines* them into two separate combinations, one a very massive particle (the sandy handkerchief) that we call the Z (or Z^0) and another which is *completely massless* (the absolutely clean one), namely our old friend the *photon* — the *electromagnetic* U(1) symmetry emerges from the magic unbroken. Indeed, the Higgs field is an extraordinary, amazingly adroit symmetry breaker!

So, summarizing the Higgs magic: the interactions are *unchanged* by local SU(2) × U(1) transformations, but the physical states *do not* have that symmetry. And after all of the symmetry breaking is done, we are left with a world in which the U(1) of electromagnetism and the SU(3) of color are unbroken. Thus the exact symmetry of our world is $SU_c(3) \times U_{EM}(1)$, with color and charge conserved in all interactions.

The Higgs' magic appears to be so powerful that perhaps he has done another mass trick as well. We have no idea why the quarks and leptons have their various different masses. Perhaps the source of their masses is also a result of the Higgs symmetry-breaking shadow dance. Again, we can construct the mathematical model that accomplishes the task. Of course, the interactions of the quarks and leptons with the Higgs would each have to be of different strength in order to yield their different respective mass values. WAIT! Why do those interactions have *different* strengths? No one knows.

Remember how we reached out for the Higgs magician but could not touch him. So far, the Higgs *particle* that we should be able to produce from the Higgs field has not been found in any experiments. Maybe we have not done the right experiments yet; maybe the Higgs is very massive and we need more energetic experiments to find it. But maybe, just maybe, the symmetry breaker is not the Higgs. Maybe the magician only appears, in the dense fog of our ignorance, to be the Higgs.

2. THE GLORIOUS GAUGE THEORY — THE STANDARD MODEL

*"Put it all together it spells..."**

After groping our way through the mist, we reach the sunlight above and beyond the swamp. The magically (Higgs) broken electroweak gauge symmetry $SU_L(2) \times U(1)$ reigns supreme here. This climb, combined with our QCD adventure, has taken us all the way through *the Standard Model* — the glorious gauge theory that has been verified to within 1% in very many different kinds of experiments. And deep within the Standard Model slumbers an explanation of the equality of the number of quark and number of lepton generations (using the basic unit of charge e for both leptons and quarks, as we have been doing all along).

*The line reads "Put it all together it spells *mother*" from the song "Mother" by George M. Cohan.

As we stand on the electroweak and the quantum chromodynamics mountain peaks (allowing me a little magic of my own, so that we can be on both at once), we realize that the Standard Model is indeed much more than a model; it is a full-fledged physical theory! It contains 12 gauge particles — the Ws, Z and photon (adding to 4) of the electroweak theory and the 8 gluons of quantum chromodynamics. The matter particles number 15 in each generation — in the first generation: the *up* and *down* quarks, in three colors, both left-handed and right-handed ($2 \times 3 \times 2 = 12$), as well as the *electron*, both left-handed and right-handed (2), and the left-handed *electron-neutrino* (1) — along with their 15 antiparticles.*

<center>Ode to the Standard Model</center>

<center>(inspired by Beethoven's music to "Ode to Joy")</center>

<center>High atop these glorious mountains
overwhelmed by melodies,
here we stand and gaze about us
bathed in heavenly harmonies.</center>

<center>Standard Model, oh Standard Theory,
we're so bewitched by your beauty,
by the splendor that we hear 'n see —
nature's wondrous rhapsody.</center>

<center>From these heights we peer yet higher
curiosity wandering free
trying through the mists that hover
t' penetrate more mystery.</center>

<center>Standard Model, oh Standard Theory,
we're so bewitched by your beauty,
by the splendor that we hear 'n see —
nature's wondrous rhapsody.</center>

In the future, when we refer to this magnificent gauge theory underlying all of nature, we will sometimes call it: $SU_c(3) \times SU_L(2) \times U_Y(1)$, where the sub-c stands for color and the sub-Y for hypercharge (usually omitting the subscripts), or by its nickname: *the 3-2-1 theory*.

Unanswered questions

- The glorious Standard-Model theory [$SU_c(3) \times SU_L(2) \times U_Y(1)$], that breaks via the Higgs magic, is not the *simple* gauge group of our original aspirations — it has those "×"s in its name! It is composed of three separate gauge groups *pasted* together? This is certainly not a unified theory. Could it be that the Standard-Model theory results from another Higgs-breaking of a larger *simple* theory — a

*For the second generation the quarks are *charm* and *strange* and the leptons are *mu* and *mu-neutrino*; for the third generation the quarks are *top* and *bottom* and the leptons are *tau* and *tau-neutrino*.

grand unified theory? There are several proposed candidate theories that will do the job, as we shall see when we peer upward.

- The Standard Model has no explanation for the equality of the fundamental constant of quark electric charge (recall that their charges are 2/3 and −1/3 of a fundamental constant that we call e) and lepton electric charge (also called e). We know this exact equality of charge to be so from our experience in the world and direct experiments on atoms, where the charges on the electrons around the nucleus *exactly* cancel the charges of the protons inside the nucleus. Our world would be a completely different place if that cancellation weren't exact! Well, be assured that just about every grand-unified extension of the Standard Model does explain that equality.

- Why are there *three* generations of fundamental matter particles? Yes, using the same e for both quarks and leptons, the Standard Model requires that they have the *same* number of generations, but why not only one, or two, or four or more?

- When a (real or virtual) W$^+$ is emitted by an isospin up quark (u, c or t), the quark changes into another that has isospin down. However, there are three generations of quarks; therefore, there are three possible flavors of the resulting isospin-down quark — d, s, or b. The final quark may be in a *different* generation than the original quark! Likewise, when a (real or virtual) W$^-$ is emitted by an isospin-down quark; the resulting isospin-up quark may be in a different generation.

 (Absorption of a W goes the other way, also with the possibility of a generation jump.) However, there is *absolutely no generation mixing* of the leptons when *they* interact with Ws; the emission or absorption of a W links electrons only with electron-neutrinos and vice versa; likewise, for the other two generations. This *mixing of quark generations* (in the W interactions) but *not of lepton generations* is an unsolved mystery of nature.

- The size of the CP (and consequently T) symmetry breaking is not determined by the Standard-Model theory, but there is a way of incorporating that breaking into the theory by specifying one number. In the next few years, experiments will determine whether the Standard Model *can* account for the size of the CP violation observed in experiments. If it can, we will still ask, what determines that number? Of course, if it does not, the experiments will have found evidence *beyond* the Standard Model!

- The universe contains very little antimatter. To account for this large excess of matter over antimatter, within the context of the Big-Bang cosmological theory, requires a CP violation beyond what could exist within the Standard Model. However, there *are* several extensions of the Standard Model that can easily remedy this deficiency.

- There are several enigmas that arise from our presumed ubiquitous Higgs field. The most prominent one (from where we stand) is the question of the energy contained in the all-pervading Higgs field. Einstein's theory of gravity tells us

that an attraction occurs between regions of space possessing field energy, just like the gravitational attraction between masses. The energy in a *ubiquitous* Higgs field would have caused space to be no bigger than a football because of the gravitational attraction among all areas of space. . .unless there is another part of gravity (whose possible existence, Einstein considered long, long ago), that just exactly cancels out the Higgs field's attraction. And we mean *exactly*! Can the effects of nature be so finely tuned that a major effect of two (seemingly) completely different interactions exactly cancel? Perhaps, such a fine-tuning is pointing us towards a broader more inclusive theory that *automatically* contains that connection between gravitation and Higgs magic? Or perhaps the magician is not the Higgs field after all!

- The laws of gravitation appear to be *incompatible* with quantum theory — it is impossible to construct a *renormalizable* fundamental quantum field theory of gravity. Is there a grander structure, for which these gauge theories are merely *effective* theories, that includes gravitation consistently? We have already pointed out that effective theories need not be renormalizable to make sense. However, it is interesting to note that the Standard Model's electroweak theory, with all of its experimental success, was found by *demanding* renormalizability.

- How do the fundamental matter particles get all those *different* masses? Is it really part of the Higgs magic? If so, why are they different and what determines their values.

- And why is there a preferred handedness? Some grand-unified theories address this question.

When we count all of the unexplained numbers, technically called "free parameters" — 11 masses (6 quarks, 3 leptons, W and the Higgs),* the 3 "charges" or strengths associated with the three simple groups that are pasted together (SU(3) × SU(2) × U(1)), the hidden assumption that $e_{quark}/e_{lepton} = 1$ (expressing the equality of the basic unit of charge for quarks to that for leptons), along with 3 generation mixings of the quarks in the weak interaction and the size of the breaking of time reversal invariance — we find that there are *19 unexplained numbers* in the (unextended or *minimal*) Standard-Model theory.† We might even wonder why leptons are not mixed by the weak interactions, which would amount to three more mixing numbers and another that breaks time reversal invariance — a total of 4 that all appear to be zero. But people are kind to the Standard Model and don't include them in the count.

The Standard Model can be extended to account for the evidence that is growing concerning the existence of neutrino mass, which we are about to explore, but we

*The Standard Model contains a (Higgs) connection between the W and Z masses, so we only include one of them as an unknown parameter.

†Note to the cognoscenti: All of these numbers can be expressed as dimensionless quantities when they are suitably defined, and so they are not part of the disagreement about the relevance of dimensionful quantities like c and h.

may wonder whether a theory with at least 19 (or 23, or more) unexplained numbers can be the *fundamental underlying* scheme of the universe? What do you think? Your reaction to this question is as valid as anyone else's. How many numbers do we expect to have to put into a fundamental theory? Would *one* be OK? "None at all" looks impossible, something like a mathematics with no axioms. So maybe *one* should be expected. Well then, how about *two*? If *two* is acceptable, why not *three*; where do we draw the line? Why do we draw the line there? This is metaphysics — very far off of our itinerary.

Now we see before us sharp images (experimental evidence) that pull us beyond the minimal Standard Model. They show us features of neutrinos that are not contained in that theory. And peering upward and beyond from atop these Standard Model peaks, we are attempting to see if these mountains join into one grand unified peak at higher elevations. From this vantage point we see a thick cloud-like mist behind them extending upward and obscuring our view. Shapes can be made out in these low-hanging clouds that stimulate the imagination, but there is no agreement as to their nature. Although no one has succeeded yet in exploring there, it is fascinating to investigate the conjectures that have been made.

Chapter 17

ABOVE AND BEYOND

1. NEUTRINO MASS AND "OSCILLATIONS"

Well here we are, more than 70 years beyond the introduction of the concept and we still have not nailed down the properties of these elusive particles. Within the Standard Model, the neutrinos are massless left-handed particles in a natural way. Massless particles have that property — they can exist with one handedness (with the opposite handedness for their antiparticles). However, experimental evidence is accumulating that neutrinos are not those simple *massless* objects of the Standard Model, instead possessing such tiny masses that they have seemed massless in β-decay experiments.

The evidence that neutrinos are not massless arises indirectly from a phenomenon called "neutrino oscillations." We can explore the nature of this phenomenon by considering a simplified model that has only two generations of neutrinos — the electron-neutrino and the muon-neutrino. Suppose that the wave function of the electron-neutrino is an *in-phase* combination (sum) of two different underlying waves that we shall call "ur-waves,"* each with a *different* mass. And similarly, the muon-neutrino's wavefunction is the opposite combination of the same two waves, i.e., they are completely *out of phase* with one another. Now we have learned earlier in our journey that the frequency of oscillation of a *quantum* wave is directly related to the energy, and hence the *mass*, of the particle being described — different masses imply *different frequencies*. So, if we *start with an electron-neutrino* and watch it for a while, the two ur-waves in its wavefunction will get out of phase — *it will become part muon-neutrino* and will have a non-zero probability of showing up in our observations *as* a muon-neutrino. In experiments involving the detection of electron-neutrinos, we would expect a correspondingly *smaller* number of them to be detected.

*We are using the prefix "ur" to mean "primordial," a meaning that comes to English from the German.

A generalization of the mathematical argument can be extended to all three generations of neutrinos, and we express this by saying that *the neutrinos can oscillate into one another (and perhaps even into types as yet undetected).* "So what does this have to do with the price of oil?" Aha — a good question! What *fuels* our interest in this mathematical conjecture is *real data*:

- A neutrino detector in a mine in Kamioka Japan* has found many more of the mu-neutrinos produced in the atmosphere above their location (from decays of pions), than from the atmosphere above the other side of Earth. (Yes, most all of them can travel right through the Earth without interacting!) Presumably, although very few would interact with anything during their long trip through the Earth, those from the other side of the Earth *oscillated* (partly) into an undetected type.

- Detectors looking at the electron-neutrinos coming from the sun have found *many less* than expected according to our present-day understanding of stars like the sun. So, perhaps some of the electron-neutrinos coming out of the sun *have* oscillated into the other two generations of neutrinos. This would indeed result in fewer electron-neutrinos arriving at our detectors, and our stellar theories would remain consistent with that data.

- In Sudbury, Ontario (Canada) a newly-built detector† is able to distinguish between electron-neutrinos and other types. Combining their results with the observations of other neutrino observatories (most notably Super-Kamiokande in Japan), the researchers in that collaboration have surmised that the answer to *the solar neutrino puzzle* is neutrino oscillations.

- By the end of 2002 as this goes to press, observations of the oscillations of neutrino beams emerging from reactors have been reported.

The mass differences among the ur-waves that could explain these results are indeed tiny values that would not have been detected in any of the experiments done before.

There are several independent experiments planned, at different locations on the Earth, to measure the content of a neutrino beam produced at a high-energy accelerator, and then measure it again after the beam has traveled hundreds of miles (or kilometers, if you're so inclined). These experiments will have the decisiveness lacking in the experiments that cannot control or measure at the source of the neutrinos.

*run by a collaboration of physicists from Japan and the US. Its present incarnation is called "Super-Kamiokande." Unfortunately, a recent accident (11/01) has severely damaged this facility.
†built and run by scientists from Canada, the US, and the UK.

Neutrino Kabuki Dance

Pervading all the world we know,
neutrino ur-waves' 'xotic dance.
Lithe swaying limbs move to and fro
as all who watch, they so entrance.

And each of them does undulate
thus changing phase, while dancing so.
But each one has its own change *rate*,
so out of sync their phases go.

Yes, each one has its own change rate
that, in its twirling, speaks of mass
so tiny — yet it does dictate
the crucial tempo of each pass.

Kabuki song, Kabuki dance
mysterious and piercing sound,
bright-colored costumes' radiance.
Phase-fans all flutter, winding 'round.

Though we can't see each one alone,
as we must look through quantum veil,
the changing flavors that have shown
proclaim the "oscillation" tale.

Now let us consider the handedness ramification of the evidence for neutrino mass. It is well known that *both* handednesses of particles that have mass (no matter how small) *must* exist. So, where is the right-handed neutrino? Why have we never seen one? Well, the neutrino has only weak interactions, which affects only left-handed particles, so we cannot detect right-handed neutrinos *that* way. Furthermore, it is the only fundamental particle that has no electric charge or color (charge), so it does not interact any other way except gravitationally — an astonishingly weak interaction, so far undetectable on the individual particle level. So right-handed neutrinos may very well exist, but beyond our detection ability.

As we step away from the *minimal* Standard Model, in which neutrino masses are exactly zero, we are puzzled by the minuteness of the masses of the left-handed neutrinos compared to the masses of the corresponding charged leptons (electron, muon or tauon). However, an elegant mathematical model exists which appears to produce extremely tiny masses for them in a natural way. It is based on the very fact that the neutrino has no electric charge or color. Since the $SU(2) \times U(1)$ symmetry of the electroweak theory has been broken, I_z and Y are no longer significant, so we can say that the neutrinos are completely neutral particles (the only neutral fundamental matter particles). Now a completely neutral particle can be the *same* as its antiparticle. For particles with spin $\frac{1}{2}$, like the neutrino, these are called Majorana (the 'j' pronounced 'y') after the creator and developer of the concept. Well, a Majorana neutrino would blow away the whole idea of lepton number! For if lepton number is not conserved then it has no scientific value, and its conservation

requires the neutrino to have $+1$ for its value to the antineutrino's -1 — they would not be the *same* particle. But there is no compelling reason to expect conservation of lepton number; it has no connection to any symmetry or other property of matter. By treating neutrinos and anti-neutrinos as identical particles, the mathematical model we have mentioned yields tiny masses for our left-handed friends (and their right-handed antiparticles) *naturally*, without putting contrived tiny mass numbers in. Within this mathematical structure, the lower their masses the higher the masses of the right-handed neutrinos — very much like a seesaw. It is consequently known as "the seesaw mechanism."

2. GRAND UNIFIED THEORIES (GUTS)

And what about the quest for a simple, elegant *unified* theory, from which all of the physical laws of nature flow? Well the $SU(3) \times SU(2) \times U(1)$-structure of the Standard-Model theory shows explicitly that it is not a unified theory — there are three separate theories "pasted" together by \timess. But this is reminiscent of the flavor-classification symmetries we visited below, that started as FLAVOR-$SU(2) \times U(1)$ and were expanded to a unified FLAVOR-$SU(3)$ theory.*

The smallest grand unified theory — GUT-SU(5)

The smallest mathematically consistent *simple* group (no \timess in its name) containing the Standard Model entwined within is $SU(5)$. The grand unified theory based on the mathematics of $SU(5)$ contains 24 generators (and their corresponding gauge particles); twelve of them are those of the Standard Model and twelve are new ones, so far undetected.

Each *simple* theory contains only one basic strength (analogous to the charge e in electrodynamics), so that the $SU(3) \times SU(2) \times U(1)$ of the Standard-Model theory contains three (called g_3, g_2, and g_1, respectively), whereas $SU(5)$ contains only one (called g_5). Well, how could a theory with only one basic interaction strength replace the Standard Model that has three quite disparate basic strengths? Furthermore, why have we not detected any effects of those extra twelve gauge particles in any of the many many experiments done in the past century? The answers to these questions are essential to our acceptance of any grand-unified theory as viable! These questions can be understood to result from our general requirement of *all* theories, referred to as *the correspondence principle* — a newly proposed theory must reduce to the theory it is meant to replace, for those situations in which the older theory was valid. So how does one basic strength (g_5) yield three (g_3, g_2, and g_1)? And why are twelve of the GUT-$SU(5)$ gauge particles so well-hidden from us?

*Recall that those were not gauge symmetries; they had only global transformations. There are no gauge particles or interactions associated with them.

Three strengths become one as masses soar

When we first entered the realm of gauge theories, we talked about the necessity of renormalizability for fundamental quantum field theories. There we saw that charge renormalization, which renders the charge uncalculable, unveils the *changes* of effective charge (or interaction strength, as we have been calling it here) as the momentum of a probing particle increases, i.e., as the energy of the scattering increases. Examining these changes of the disparate interaction strengths of the Standard Model, it has been shown that at the fantastically high scattering energy that is more than one thousand trillion (a 1 followed by fifteen zeros or 10^{15}) times the mass energy of a proton, the three interaction strengths (referred to as g_3, g_2, and g_1) all would become approximately the same value. (If the world had a property called *supersymmetry*, which we will discuss later, we could even remove the word "approximately." For the moment we restrict our discussion to the features of ordinary GUT-SU(5).) But the higher the energy of scattering the closer the distance being probed, so that the distance at which the 3-2-1 symmetry behaves like SU(5) is very tiny indeed — about a thousandth of a trillionth (10^{-15}) of the size of a nucleus or about 10^{-28} centimeters. We don't expect to ever be able to measure that close!

However (following the Big-Bang scenario), we find that at a very early time, when the universe was very hot, the average momentum of particles was indeed high enough for the three interaction strengths of the Standard Model to have been the same value. We can legitimately propose that *earlier* in the universe's existence, when the temperature was even higher, the symmetry that reigned was SU(5) with only one interaction strength (g_5). Somehow, as the universe went below the temperature at which the three interaction strengths of the Standard Model are the same, the SU(5) symmetry must have broken in such a way that those three interaction strengths became unequal as the temperature dropped further. And we can imagine that the breaking of the GUT-SU(5) was accomplished by a little *grand-Higgs* magic at that very high temperature; only the subgroup SU(3)×SU(2)×U(1), which is the Standard-Model theory, remained unbroken at that time (all the way down to much lower temperatures).

And what of the extra twelve gauge particles of SU(5)? Well, they each acquired a mass that roughly corresponds to the fantastically high energy at which the symmetry breaking of SU(5) occurred (about 10^{15} times the mass of the proton). This is analogous to the Higgs-mechanism acquisition of mass by the Ws and Z at the much much lower energies where the electroweak theory is broken. Effects due to those twelve enormously heavy gauge particles of broken SU(5) could not possibly have been detected by *our* experiments (so far) that are at energies so very much smaller. This *grand-Higgs* spontaneous symmetry-breaking would occur at an unimaginably high temperature (about ten thousand trillion trillion degrees Centigrade or 10^{28}C), which is believed to have been the temperature of the universe very very shortly (about a tenth of a trillionth of a trillionth of a trillionth of a

second or 10^{-37} seconds) after it was created. So, the scenario, for an unimaginably short time interval during the very early universe, consisted of quarks and leptons undergoing SU(5)-symmetric interactions with the SU(5) gauge particles. Then, almost immediately, that symmetry broke to leave the Standard-Model theory's SU(3) × SU(2) × U(1) still standing (to be broken at a much, much lower energy and temperature).

More GUTs

Although it would make good copy, I cannot say that there are as many grand unified theories as there are theorists studying them. This is by no means true. Rather, I can say that there are several viable grand unified theories besides SU(5). A particularly interesting one is SO(10).

The gauge group SO(10) has some very attractive features. We have discussed the groupings of particles into multiplets of symmetries earlier in our travels (e.g. the isospin doublets of quarks and those of leptons). SO(10), like all of the grand unified theories, treats each generation of quarks and leptons separately and identically; therefore, we need only look at its treatment of the first generation. In SO(10), all of the first-generation left-handed fundamental matter particles are in *one* multiplet — a 16-plet, and all of the right-handed ones are in the corresponding anti-16-plet. Well, how many first-generation *left-handed* fundamental matter particles have we met? There are 12 quarks (u, d and \bar{u}, \bar{d} in three possible colors: 4 *times* 3) and 3 leptons (the electron, the positron, and the neutrino — no left-handed antineutrino has ever been detected). That adds to 15. Likewise, all quarks and antiquarks come in right-handed versions, along with the electron, positron and antineutrino (no right-handed neutrino has ever been detected). So the total is 15 of each handedness, but the 16-plet and anti-16-plet of SO(10) each have one more! The extra member in each of these multiplets corresponds to the missing handedness of the neutrino, namely a right-handed neutrino and the corresponding left-handed antineutrino. In fact, such particles would *have* to exist if the neutrino oscillations visited in the previous section are really occurring — and that experimental evidence is mounting!

An interesting explanation of the observed seemingly *ad hoc* handedness of nature emerges from this theory. For in *unbroken* SO(10) left-handedness and right-handedness have equal footing — there is no preferred handedness here. The handedness that we see in nature is proposed to be the result of another magical Higgs symmetry-breaking at huge energies — we could say that this renders the left-handedness we have seen (at our "low" energies) *natural*, in the sense that we would not expect breaking of the two handednesses at identical energies. An important result of this high-energy symmetry breaking is the acquisition of *very large masses* also by the gauge particles that interact with the right-handed neutrino and the corresponding left-handed antineutrino. Such large masses would account for the fact that we have not detected any effects of those gauge particles.

Well, SO(10) has almost twice as many generators as SU(5) so it is much more complicated, and most other suggested GUTS are larger still. They all remain purely conjectural at this time.

Proton decay?

The proton is the lightest baryon, so that if it decays (necessarily into *lighter* particles), the decay products must be non-baryonic matter. Uh, oh! There goes baryon number conservation! Not to worry! Conservation of baryon number (just like lepton number) is not connected with any symmetry; it may just be an idiosyncrasy of the Standard Model. Most *grand unified* theories, and in particular SU(5), have multiplets containing *both* quarks and leptons. (We just saw an extreme example in SO(10), where *all* of the left-handed quarks and leptons of the same generation are in the *same* 16-plet and likewise for the right-handed matter particles.) This means that the symmetry operations of the group mix quarks and leptons together and one type *can* change into the other when interacting with the appropriate gauge particle. As a result, in these grand unified theories the proton can decay several different ways, for example into a π^0 and positron — $p \rightarrow \pi^0 + e^+$,* with the baryon number dropping by 1 [going from 1 to 0] as does the lepton number [going from 0 to -1]. (In fact, for all baryon-number-changing events in SU(5) *as B goes, so goes L!*)

The GUT-SU(5) predicts that a proton has a probability of $\frac{1}{2}$ to decay during a time interval of about ten million trillion trillion (or 10^{31}) years. For large numbers of protons, this is the time it takes for about half of them to have decayed; it is often referred to as the *half-life*. Well that *seems* OK since we believe that the age of our universe is only about fourteen billion years, so we (and our protons) *can* be here to consider the SU(5) GUT. The probability of any *particular* proton decaying this year is very, very, . . . , very small! But suppose we watched about a billion trillion trillion (or 10^{33}) protons. We *should* see some of them decay this year.

So how do you watch so many protons at once? It is not difficult to imagine because the ordinary matter around us has an enormous number of molecules, with quite a few protons in each one. A few thousand tons of material, water or iron, or etc., would do! Well, it is actually no simple task to build and run such an experiment but, at about a dozen locations around the world, large amounts of material *were* scrupulously watched (by state-of-the-art equipment) for proton decays. The data showed no evidence of proton decay; the probability that it decays at all is therefore less than 1% of the extraordinarily tiny probability predicted by GUT-SU(5) and may be exactly zero. This is summarized by the statement: The half-life of the proton is *at least* 10^{33} years. And what about all of the expense involved. Do not be concerned! Many of these sites have been converted into state-of-the-art

*At the quark level, the two u quarks in the proton (uud) fuse to produce a virtual version of one of those extremely heavy gauge particles that then decays into a \bar{d} plus a positron (e^+). The \bar{d} emerges bound to the left-over d that was in the proton to form the outgoing π^0.

neutrino detector facilities, carrying out cutting-edge astrophysical observations for about ten years now. They detected the neutrinos from the supernova that occurred in 1987, thus confirming part of our understanding of the behavior of stars, and have recently found important evidence of the neutrino oscillations we have already discussed.

Well, the proton decay data, none observed, certainly proclaims the death of GUT-SU(5). But we shall see that the extended *supersymmetric* GUT-SU(5) still lives!

3. THE EARLY UNIVERSE

The Big Bang

The name "Big Bang" was coined in 1950 to ridicule that theory of the origin of the universe, by a proponent of the competing steady-state theory.* But much evidence has been found since then that leads most scientists to believe that the universe we know did indeed begin that way. This theory proposes that there was an infinite density of energy (corresponding to an infinite temperature) located at a space-time *point* that was the *whole universe at the beginning of time*. It expanded very rapidly, like the surface of a rapidly inflating balloon, and began producing particles (or strings, if you are of that persuasion) as it cooled down. Most of those particles were moving at very high speeds, corresponding to the very high temperatures. The subsequent history of our universe led to *us*. (The jury is out as to whether all of this development was worthwhile.) That story, as well as the experimental evidence, is far from our trails and we shall not pursue it. Perhaps your curiosity about it will take you on some intriguing journeys in the future.

And so we believe that, for an unimaginably short time just after the Big Bang, the universe itself produced particles at speeds much, much higher than could the highest energy particle-accelerator of our dreams — there is no way we could even *envision* building anything that could produce such enormous energies!

Quark-gluon plasma

Were the quarks and gluons locked inside hadron clumps in the very early universe, or were they moving about relatively freely as that quark-gluon soup called "a quark-gluon plasma"? That is an interesting question. There is a new collider located on Long Island that will perhaps provide some insight for us. It produces collisions between the nuclei of very heavy atoms. The very high energies of the colliding particles, along with the fact that the collision squeezes them into a very tiny region of space, might, just for an instant in that very small place, reproduce

*In fact, the magazine *Sky and Telescope* had a contest in the 1990s asking for suggestions of a better name, but none was chosen.

the conditions of the very early universe. For a fleeting moment at that location we may see evidence of free quarks and gluons — *colors unfurled!* That would be quite a feat. The particle collider is called the Relativistic (emphasizing the high speeds) Heavy Ion (an atom missing one or more electrons[†]) Collider — RHIC is the acronym. So, as the new century progresses, watch your newspapers for the results from RHIC.

4. SUPERDREAMS

The superdreamers

Now we meet the superdreamers, some of whose previous dreams have been dramatically verified by experiment and are already incorporated in our theories; those dreams have made our journey to this place (right to the Standard Model) possible. Let us listen to the new dreams they have created, based on their physical intuitions and the shadowy shapes they perceive in the heavy mist before us.

Supersymmetry — the deliverer

We are being urged by the superdreamers to stare into the mist. There seems to be much more structure than we first perceived. The facets of the structure (referring metaphorically to the *mathematical* structure) appear to be emitting a fascinating glow, which hurts our eyes as we stare. It is the glowing structure of *supersymmetry*. And now, as our eyes grow used to the eerie light, we can see the main features of this apparition.

The mathematics of supersymmetry eluded physicists until the 1970s, when it was finally realized that bosons and fermions (whose behaviors are completely different) can be in the *same* multiplet, for a bizarre type of symmetry group. Because theories with that unusual feature reach beyond previous concepts of symmetry, they are referred to as *supersymmetries* (SUSYs). We have already mentioned that the SUSY version of SU(5) is a viable grand unified theory that has not been ruled out by experiment. Supersymmetric theories have the weird feature that, for every fundamental particle we know (and love), there is another that is identical except that its spin is shifted down by $\frac{1}{2}$: Besides the electron (of spin $\frac{1}{2}$) there is a corresponding particle of spin 0 — the scalar electron called *selectron*; besides the quarks (of spin $\frac{1}{2}$) there are corresponding particles of spin 0 — scalar quarks called *squarks*; and likewise there are the *sneutrinos*. The s prefix stands for the word *scalar*, the name used for rotationally invariant objects — exactly the behavior of spin 0 particles. However, when talking about the scalar u or scalar d quarks, *we do not* abbreviate to *su* or *sd*![*] So watch your s's!

[†]It has a net positive charge so that it is easily accelerated.
[*]The s would be misconstrued to represent a strange quark!

For the gauge particles, the naming scheme has a different tone — no s's in sight! The gauge particles have spin 1 (except for the graviton, which we will look at separately), and their spin partners have spin $\frac{1}{2}$. Taking a cue from the neutrino, the lexicon creators attached the suffix *ino* to the names — for the photon, the spin $\frac{1}{2}$ partner is the *photino*; for the Z, the spin $\frac{1}{2}$ partner is the *Zino*; for the Ws, the spin $\frac{1}{2}$ partners are the *Winos* (not the way it looks, it is pronounced to rhyme with neutrino); in general, the spin $\frac{1}{2}$ partners of the gauge particles are called *gauginos*. The spin 2 graviton also has a partner, of spin $3/2$, called the *gravitino*.

Supersymmetric grand unified theories provide a rationale for the possible existence of a *relatively light* fundamental scalar particle, the anxiously sought Higgs particle of the electroweak theory. We expect the mass of that particle to be no greater than about ten thousand (10^4) times the proton mass, otherwise all of the Standard-Model calculations, which have been so carefully verified by experiments, would be wrong. The *existence* of the supersymmetric partners of the particles we know *prevents* the interactions from causing the Higgs to have a (renormalized) mass comparable to those of the extra (enormously heavy) gauge particles of the grand unified theory — about one thousand trillion (10^{15}) times the proton mass, which is one hundred billion (10^{11}) times too large!

But what about proton decay? Is this déjà vu — all over again? Not at all! Instead, this is "cannot view." The viable supersymmetric theories predict the lifetime of the proton to be about 10,000 (10^4) times longer than the gigantic experiments that watched protons could have detected. Whew! Grand unified symmetry has been saved by supersymmetry — as of now. Supersymmetric SU(5) lives! And don't count out larger supersymmetric theories either.

Where have all the partners gone?

So where are all of these spin partners of the fundamental particles we *know*? Why have we not seen any effects from them? Well, they are not around because the Higgs magician has done it again — broken the supersymmetry without a word to anyone! This means that the masses of the *superpartners* of the ordinary particles we know should be very large \cdots but \cdots there is a possibility that they are not so large as to be out of reach. The Superconducting SuperCollider (SSC), that was canceled by Congress in mid-construction, would have been able to answer this question. Perhaps the smaller European project (called the Large Hadron Collider (LHC)) that is being constructed in Geneva will dispel enough mist to provide some answers. Or perhaps careful evaluation of data that has already been recorded will show evidence of their existence. *TO BE CONTINUED!*

Supergravity (SUGRA)

Is there really something better than gravity? No, no, not that, but it's "super" just the same. We have seen, before our recent climbs, that continuous symmetries

can be extended to gauge theories by including transformations that are *different at different points in physical space-time.* When this is done to the supersymmetric theories, they become *local* supersymmetric theories and, to the astonishment of all who see the amazing result, they contain Einstein's well-known (and well-verified) theory of gravity, called general relativity, *automatically.* Well the lack of renormalizability remains a problem (probably indicating that this can only be an *effective* theory) but the beautiful manner in which gravity is automatically embedded in the gauged supersymmetric theories is indeed astounding.

Superstrings

Yes the supergravity theories that seem to loom up in the distance are not renormalizable. However, some of our superdreamers are dreaming even more expansive dreams that contain supergravity, which contains supersymmetry, which contains a grand-unified field theory, which contains the Standard Model, as successive effective-theory approximations of the *actual* master plan. As effective theories they do not have to be renormalizable! In those dreams, the fundamental entities in nature are not the point particles (quarks, leptons and gauge particles) whose existences have been well verified; they are not point particles at all. Instead, the fundamental entities of nature are *superstrings*, tinier than any strings you can imagine — about 10^{-33} cm, called a "Planck length." (Modern-day accelerators can probe only to about 10^{-16} cm.) They are one-dimensional like an ideal string and not made of anything else (the way ordinary strings are composed of molecules). Their sizes are so incredibly tiny that they appear as point particles to us (with the usual quantum mechanical wavefunctions, etc.).* In their different guises, they behave as quarks, leptons or gauge particles. String theories are automatically renormalizable because of the finite (albeit incredibly minute) size of the fundamental strings.

One peculiarity of string theories is that they appear to require the space we live in to be 9 (and in some scenarios 10) dimensional. But we know of only 3 space dimensions. So where are these many extra dimensions? Our resourceful string theorists have managed to create mathematical scenarios in which these extra dimensions do not stretch out to infinity as the 3 we are aware of do. Instead they close on themselves the way a circle or a sphere does, and their "circumferences" or "surfaces" are so extraordinarily small (roughly the sizes of strings themselves) that we could not have noticed them so far. Nor have any experiments been devised so far that could reveal the presence of such extra dimensions. But the mathematics of these imaginative theories certainly requires their existence. Despite this weirdness, it is possible that these fundamental strings do form nature's grand-viol, with the vibration that that *supermusician* sets up on each determining what particle (quark,

*String theory has even led to the consideration of two-dimensional membranes, higher-dimensional fundamental objects called p-branes (not 'pea brains'), and something called M-theory.

lepton, or gaugon) it appears to be. Neither renormalization inconsistencies nor
any other flaws would be present in this, nature's grand musical instrument. When
all is said and done, the superstrings of the grand-viol may indeed be as real as
our concept of *water*. (We all know that the rain in Spain* falls mainly in the
four-dimensional spacetime continuum.) However, the theories of strings and of
superstrings have no foundation of experimental observation at all; they are still
in the category of *dreams*, albeit sophisticated mathematical dreams, and may be
merely figments of the superimaginations of our superdreamers.

5. LARGE EXTRA DIMENSIONS

There are those who actually see *large* extra space dimensions in the mists before us.[†]
They attribute to them the *apparent* weakness of gravity and the unusually large
differences that we find among the masses of the 'so-called' fundamental particles
and among the strengths of the different interactions. In their scenarios we live
on a 3 dimensional "brane" (taken from the word "membrane") within a larger
dimensional space. Some of those scenarios even include other 3 dimensional branes,
separated from us by spans in the extra dimensions, that influence the 3D brane we
live on. They are something like parallel universes, although the "laws" of nature
may be very different in them. These musings, which are evolving as we speak, do
actually make predictions that could be tested *experimentally* in the next few years,
so we all await the near future with (not quite) baited breath.

6. HAVE WE REACHED BOTTOM?

And now we end where we began, asking the same question. Are the particles we
consider fundamental actually composites of more fundamental entities? Indeed,
string theories are already offering one possibility, and there are less ambitious
proposals with fanciful names like "preons" and "rishons" as the entities of a more
fundamental level. Well, nowhere in the *research* literature are the particles we have
treated and referred to as "fundamental" officially named that! That adjective has
been used *provisionally*, in order to distinguish them from the slew of "elementary
particles" (found in the last half of the twentieth century) most of which are com-
posites — the hadrons. Perhaps, instead we should have stuck to the seductive and
defensive *whimsy*. Could we not have called these quarks and leptons "the nuts
and raisins" of matter and the particles of the interactions "dough lumps?" Those
names would certainly fit with *color*, *flavor*, and *glueball*, as well as the other whim-
sical names. But, to wax serious for a moment, let us not forget that despite those
names, these particles and their interactions have been studied as carefully as our

*With apologies to Lerner and Loewe, the creators of the Broadway hit "My Fair Lady."
[†]Recently, string theories have been proposed that have *large* extra dimensions also.

hard science can possibly do. The question of fundamentality will always be with us, and its intriguing siren call will surely entrance the scientists of generations to come.

A RECONSTRUCTION

OF

THE ORIGIN OF THE PARTICLES

THE DESCENT OF SYMMETRY

for narrator and chorus

Cast of Characters*

The fermions — the fundamental-matter (spin-1/2) guys

quarks fermions of "color" (associated with nuclear forces)
leptons colorless fermions (like the electron)

The gaugons — particles of the interaction that shmooze with fermions
and sometimes with one another (spin-1 except for the graviton)

gluons "colorful" gaugons of nuclear forces
photon a gaugon of electromagnetism, including light
W charged gaugons of radioactivity
Z an uncharged gaugon of radioactivity
gravitons (spin-2) gaugons of gravity

The hadrons — quark clumps

mesons shy types of hadrons (discovered in the 1940s)
baryons the proud and famous *proton* and *neutron*, and their heavier cousins

Others

Higgses - trouble-making strangers
gluonium - a character made of two lovingly-entwined *gluons*

*Note to the cognoscenti: Supersymmetric partners of all the characters here may be lurking in the shadows; they do not take part in the action of our drama.

Before the ancient times, ago
before the time before,
just after time began to go
and space was little more,

the universe had symmetry
both local and so grand —
though there was no geography;
there was no sea or land.

chorus: Oh symmetry, oh symmetry,
your beauty does astound,
and all of us are filled with glee
whenever you are found.

'Twas hotter then than hot can be
'twas hotter than the sun,
and all there was, was *energy*;
the "world" had just begun.

And all was ruled by symmetry
unbroken and supreme,
and "particles" began to *be*
as in a fev'rish dream.

Some say they're strings 'va grand-viol
that nature plucks and plays.
Each different tone has its own role,
as particle 't portrays.

And so the particles we know
are superstrings, perchance.
But it was oh so long ago
that they could twist and dance.

For stiffen did those superstrings,
in universe, out-bound,
as quarks, as leptons wandering
or gaugons rushing 'round.

chorus: For all space just keeps expanding
and the temp'rature does fall.
The inev'table demanding
of the laws that govern all.

And so it was long, long ago,
the universe aboil,
that superstrings had to forgo
their craving to uncoil.

Grand symmetry, so mighty then,
its dominance proclaimed.
And all the particles we ken
were caught in its grand frames.

chorus: Oh symmetry, oh symmetry,
your beauty does astound,
and all of us are filled with glee
whenever you are found.

The fermions they had no mass
and spin one-half had they.
But diff'rences 'tween them, alas,
would soon come into play.

And many gaugons there arose
from th'interaction's play.
So many different types of those
but much alike were they.

All spin-one gaugons, full of flair,
did crave for fermions.
And some did even make a pair
of those desired ones.

These spin-one ones in *virtual* guise
'tween fermions did go.
And carried interactions' ties
from one to th'other so.

And thus they functioned in their course
as interactions' hands.
Their *masslessness*, it was the source
of th' symmetry, so grand.

chorus: Oh symmetry, oh symmetry,
your beauty does astound,
and all of us are filled with glee
whenever you are found.

As space expanded full and all,
the particles were stirred.
The temp'rature did fall and fall
and then a *change* occurred.

We're told that hidden strangers struck:
the mighty Higgses bold.
From vacuum, some *mass* they plucked
for many to enfold.

continued over

The change was sudden and extreme:
 some gaugons now had mass.
Yes, just as in a horr'ble dream
 this furor came to pass.

The symmetry was broken so,
 fragmented through and through.
But part remained to govern, though
 much more was lost, 'tis true.

Yes, symmetry survived the break
 but something wasn't right!
For then its parts, they did forsake
 equality of might.

chorus: And all space just keeps expanding
 and the temp'rature does fall.
 The inev'table demanding
 of the laws that govern all.

Now, for an instant, there remained
 a symmetry quite wide.
"Electroweak," that was its name,
 but cooling broke its pride.

Small symmetries *still* seemed to *be*;
 of diff'rent strengths were they.
But fermions' equality
 this world did not obey.

And then the quarks did so denounce
 the leptons *weak* reply.
Their *strong* outlook made them renounce
 the leptons, by and by.

But soon another time would come
 with great catastrophe.
The quarks would lose all their freedom
 while leptons remained free.

Beware those gluons, they were told.
 Beware their strength and ken.
For once they get you in their fold
 You'll not get out again.

chorus: And all space just keeps expanding
 and the temp'rature does fall.
 The inev'table demanding
 of the laws that govern all.

And bags were formed of gluon fields;
 three quarks locked in each one.
The forces that each gluon wields
 thus made the *baryon*.

And when a quark met antiquark,
 a *meson* (bound) to make,
these fermions tried hard to balk.
 Their freedom was at stake.

But finally all quarks were bound
 and antiquarks as well.
And gluons were together wound,
 their colors thus to quell.

So color passed out of our world
 with no apparent trace.
Gluonium and hadrons swirled.
 No color showed its face.

But hidden deep inside these lumps,
 within them there's no doubt,
the color force rages and romps
 and some effect leaps out.

'Tis stronger though than might be thought,
 the strongest force — deluxe.
Yet just a remnant of that caught
 in each bag's web of flux.

chorus: And all space just keeps expanding
 and the temp'rature does fall.
 The inev'table demanding
 of the laws that govern all.

And many gaugon types were gone,
 too heavy to complain.
But photons *still* do carry on,
 for massless they remain.

To bring 'lectromagnetic force
 'tween particles with charge,
a photon loves to fly the course
 o'er lengths, from small to large.

And how did all the leptons fare?
 Did they survive as well?
By shunning gluons with great care,
 they're free, though th' others fell.

continued over

But charg'd leptons are not *all* free.
To interact 's their plight,
with photons that ignore their plea
sometimes, and bind them tight.

A weak force did survive, also,
the symmetry's descent.
Its gaugons' masses rather low;
its power not all spent.

The W 'n Z, those gaugons weak,
with leptons interact.
And in the lumps, to quarks they streak
their force-role to enact.

(These weak-force gaugons look at spin
and mingle with great care.
With lefties only will they grin,
with righties they don't dare.)

And many weak decays, arranged
by Ws, occurred.
The flavors of the quarks were changed,
then they were reinterred.

Thus many hadrons did decay,
so many types to few.
The baryons' large type-array
became now only two.

The proton and the neutron, "white",
were all the kinds there left.
No quarks or gluons were in sight
Of color 'twas bereft.

And mesons, "white", were there anon,
with colors in their glue.
But they decayed, all color gone,
to lepton-photon stew.

And fourth among the forces left
without depravity,
with gravitons that link all heft,
pervaded gravity.

And now *our* story does begin,
we, *remnants of the fall*:
THE BEGINNING*

*This is *the beginning* spoken of in the first book of the Bible: B'RAYSHIYT (meaning "In The Beginning"), translated as *Genesis*.

Part IV

THE EXPERIMENTS —
EXPLORATION AND DISCOVERY

*The journey that has brought us to this high place in our understanding of our world could not have been made without the explorations and discoveries made during numerous experimental endeavors — the ingenious and insightful work of those who created and performed the experiments! Their creativity in designing and building state-of-the-art equipment and their intuitive understandings of what to look for and how to look must be proclaimed to all who journey here. They are the explorers who have probed "the very entrails of nature!"***

This very brief over-view of the experimental aspects of elementary particle physics is only meant to whet the appetite with a taste here and a taste there. (More exhaustive treatments are available in several other books, and more details about accelerators, detectors and the like are available at the appropriate Websites.)

*Francis Bacon in his *Novum Organum,* an early seventeenth century work reprinted in *The World of Physics* by Jefferson Hane Weaver, Simon and Schuster (1987).

Chapter 18

THE ROLE OF EXPERIMENTS AND EXPERIMENTALISTS

1. WHAT IS A *SCIENTIFIC* THEORY?

Experimental results form the crucial validation of a theory's integrity as a *scientific* theory. Indeed, no theory has scientific merit until it has been tested by various kinds of independent experiments. A theory that has no experimentally testable consequences is not a *scientific* theory — the more experimental predictions and kinds of predictions made and verified, the more *powerful* its scientific credentials. *Nature itself is the final arbiter* of the merits of a scientific theory.

Experimental explorations into new realms of our world have blown beautiful, long-held theories right out of the water, bringing us completely new insights about the universe. We have seen that Rutherford's discovery of the existence of the nucleus completely changed our concept of the atom, the discovery of the neutron changed our understanding of nuclear structure, and the discovery and detailed studies of all the mesons and baryons revealed enough of the structure of matter for the construction of the Standard-Model theory. Even the concept of quarks, suggested in 1964, from which that beautiful and powerful theory had been wrought, was not embraced by the high-energy physics community until 1974, when previous experimental discoveries hinting at the quarks' existence were put into sharp focus by the discovery of *charm*. Exploration and discovery have been the wellsprings of new ideas, as have conceptual and mathematical creations been the sources of new experimental tests of nature. This is *the symbiotic relationship between theory and experiment.*

2. AND WHAT IN THE WORLD IS AN EXPERIMENTALIST?

In scientific endeavors, the scientist or scientists generally have some conceptual framework and perhaps some definite mathematical predictions, and then proceed to design and perform experiments to test the validity of their notions. These

explorations of the unknown are essential to the further development of our under-
standing of nature. So what in the world is *an experimentalist*? Now we are treading
on dangerous ground! One wrong word or implication could bring an avalanche of
abuse down on this poor author.

An uncontroversial fact about the field of elementary particle physics is that
the experimental techniques and devices are so complex and so often at the cutting
edge of all technology, that it is usually a full-time career to conceptualize, design,
and build the equipment and then conceive and artfully perform the experiments.
Yes, let's not forget that it is often an *art* to make a difficult, *reluctant* experiment
actually work. The interpretation of the experimental results, with error estimates,
etc., is also no mean task, as we shall soon see.

Sometimes, an advance in technology can enable an insightful experimentalist
"to go where no one has gone before." Above all, a powerful intuition, a "gut"
understanding of the implications of our present concepts, is necessary for experi-
mental exploration and discovery; furthermore, that intuition can sometimes lead
to wonderful iconoclastic dreams beyond extant theories. Magnificent physical in-
tuition coupled with the requisite experimental expertise and artistry are the *stuff*
that lead to landmark discoveries. But if I have given you the impression that ex-
periments are somehow mystical, be assured that they are very precise, carefully
designed, performed and analyzed endeavors.

So what about the theories, the broad dreams created in the mind, and their
innovative mathematical formulations? Couldn't those completely occupy the time
of the person thus engaged? Yes indeed, and those people are referred to as "the-
orists" (or "theoreticians"). — No avalanche in sight! — Generally the theorists
do not get involved with the equipment at all. Some theorists do ascertain (mathe-
matically) what the results of experiments tell us about the proposed theories and
their models, and determine what kinds of experiments ought to be performed to
definitively test them.

So are you getting the idea of what an experimentalist is? I hope so. They
are those who are involved in the conception of experiments, the designing and
building of the equipment, the performance of the experiment, and the analysis of
the results. They often also do the kinds of dreaming and calculating that theorists
do. However, I can take the risk of an avalanche to say that the last person in this
field to create great theory *and* do crucial, beautiful experiments was Enrico Fermi
(working in the period from the 1920s to the 1950s). Do you hear any rumblings
from above?

3. THE IMPORTANCE OF CONFIRMATION OF
DISCOVERIES AND OF ERROR ESTIMATION

One of the great American physicists of the twentieth century was Professor I.I.
Rabi, who brought the quantum mechanics that had been developed in Europe

to United States universities. In a class he taught at Columbia University (on statistical mechanics), which I was fortunate enough to be able to attend, he told two stories; one definitively conveyed the crucial importance of the independent confirmation of discoveries, and the other the importance of error analysis. I will tell you these stories as I recall them these many years later (with some memory refreshing from published accounts of the first story). Despite the haziness of my memory, the essences of these stories are true and they still hit their marks!

The importance of independent confirmation

In 1903, amidst the startling new discoveries of x-rays, electrons, quantization, etc., a new and remarkable discovery called "N Rays" was made at the University of Nancy in France. A highly regarded researcher, (Professor R. Blondlot), and his graduate students had found these rays in his world-class optics laboratory. (In total, Blondlot published twenty-six articles and a book on "N Rays.") However, researchers at other state-of-the-art labs around the world could not reproduce those results. In fact, none of them saw N rays at all! About a year later, Professor R. L. Wood of Johns Hopkins University asked to visit the lab in Nancy to observe. Wood was indeed graciously invited to witness the experiment in progress, so that he could see the effect with his own eyes. This is not an uncommon courtesy to extend to a fellow researcher. Omitting some of the details in the published source* that has refreshed my memory, let me cut to the part that Professor Rabi told us. The visiting scientist entered the extremely dark room and, after waiting for his eyes to dark-adapt, saw his ebullient colleague pointing towards the eyepiece of the apparatus. The enthusiastic host and his graduate students took turns peering into that eyepiece and proclaiming "There it is," and "Yes, I see it," etc. (Whether this was in French or English, we were not told.) The visitor was then invited to look as well but, although looking as carefully and thoroughly as he could, he saw nothing — absolutely nothing!

In order to understand the next part of this story you must know a little bit about how spectra are produced. For light, a piece of glass shaped something like a wedge (called "a prism") sits in the middle of the apparatus to spread out the waves of different wavelengths — a rainbow of colors from the incoming white light. (In the setup shown to Wood, it was an aluminum prism.)

Now you may be able to guess what happened next. This is my memory of the story we were told. There, in that very dark room, the visiting scientist took the prism out of the apparatus and hid it in his pocket while no one was paying attention. However, his host and the graduate students continued to take turns looking and proclaiming that indeed they could see the effect. What a triumph for the power of suggestion! But this is not the end of the story. As I recall, Professor Rabi told us that the visiting scientist left France and returned to the US with that

*Macmillan Encyclopedia of Physics, Volume 3, 1996.

prism. He then submitted a short letter to a noted physics journal to make public the particulars that I have recounted to you. What chutzpah!

This episode was indeed a case of mass delusion. The lesson we learn from this story is that any new "discovery" should be confirmed by *independent* experiments done by different experimenters with different equipment. In fact, we should also ask what experiments have already been done that *should* have revealed this effect and, if there are any, why has it not been seen in those experiments. Sometimes there are such experiments, but the data analysis was done in such a manner that the effect would not have been revealed; if so, then *re-analyze* the data (if it is still available) so as to confirm the discovery!

Error analysis

This story took place a little further into the twentieth century. Two graduate students were working together on an experiment. One was in charge of changing the magnetic field produced by a big electromagnet by turning a dial to raise the current through its coils from one value to another. At each current setting, the other student was to read a meter that showed the strength of the effect being studied. "OK; take the reading." The data showed that at a particular setting of the magnet current the effect reached a maximum, the meter's needle being higher than for other settings, and then went down again at the next setting. They did many runs of the experiment, each ranging from low magnetic field to high. When Professor Rabi looked at their data, he immediately saw that the data was not valid. It was not that the maximum occurred at a different setting in each run, but just the opposite: the maximum occurred at *exactly* the same setting in all of the runs! And what's wrong with that? It sounds as if those guys did a careful and accurate experiment. Nope! When readings are made from meters, etc., the last significant figure is estimated. Other random effects from the surroundings would be expected as well. We should therefore see a distribution (around the actual value) that looks like a bell-shaped curve. Well, the fellow reading the meter *was* in a *different* room than was the one adjusting the magnet strength — and they are honest fellows! So how did this data get contaminated?

The experiment was redone using a non-verbal and non-visual signal to tell the meter reader when to record its reading. In this second set of runs, no maximum was observed at all! It reminds me of the story of Hans the horse, which was purported to be able to do addition. "Five plus six," anyone chosen from the audience would ask, and Hans would dutifully tap his hoof eleven times — a brilliant horse. No tricks, no signals or whispers; why his owner would surely have been able to pass a lie detector test if they had had the equipment at that time. Well, somebody asked if he could test Hans without the owner being present. "Of course," the owner said proudly (but of course he said it in German). "Four plus three," and Hans began tapping his hoof — "four, five, six, seven, eight ... wait a minute, Hans does not

know when to stop." His owner was not there for Hans to see the satisfied relaxation of his facial muscles when he reached the right number! Likewise, the meter-reading graduate student could no longer subliminally detect the unintended excitement in his collaborator's voice. Mystery solved!

Unintended suggestion occurs all the time in patient-trials of new medications. That is why double-blind trials, in which neither the doctor nor the patient is told whether the patient is receiving the new drug or just an inactive substance (called a "placebo"), are the only studies considered valid. In fact, there even is a "placebo effect" among those receiving the placebo; about 30% to 40% will experience or imagine a temporary benefit perhaps because their hopes have been raised or due to the extra attention.

The kind of random errors that we have mentioned above are collectively referred to as "statistical error." The more trials we perform the narrower the bell-shaped distribution of readings becomes; consequently, we can be more certain of what the actual reading should be. (Statistical errors are reported with a "plus-or-minus" in front of the most likely range of deviation* of the true number from the number at the center of the bell-like distribution. For example: When measuring the mass of an object, we might have found 537.2 ± 0.2 kilograms; the average reading was 537.2 and the spread around it, that most likely contains the actual value, is represented by ± 0.2. This tells us that the actual value is most likely between 537.0 and 537.4.)

Errors due to the experimental setup — systematic error

Besides the statistical error that is present in every set of measurements made, there is also the error that is inherent in the equipment employed and the manner in which it is being used. After all, there is no ideal apparatus; it must be made of *real* materials in this *real* world. The experimentalist must use his/her best judgment along with knowledge of previous experiments to estimate the amount of error that any piece of equipment might introduce. These estimates rely on the experience and intuition of the experimentalist; they are indeed part of the art of experimentation! This type of error is referred to as "systematic error." In modern-day particle experiments, the combined statistical error and the systematic error are reported separately. The care and conservative treatment that error receives in the analysis of experimental results is usually laudatory. Not only is error used in reporting data and in the subsequent comparisons with other results and with theoretical predictions, but also preliminary estimates of the systematic errors can affect the feasibility and design of proposed endeavors.

*Technically, this refers to one standard deviation.

Experimental tests of a probabilistic theory

Our present scientific understanding of nature is that we live in a quantum-universe. Quantum theory is inherently probabilistic so that repeated experiments are *not* expected to produce the same results, rather a *distribution* of values should occur. This certainly complicates the error analysis, leading to an inherent "width" of measurements that cannot be reduced further by improved measurements or increased observations. In fact, the value of the quantum-width contains important information about the interaction, as we shall see later.

Chapter 19

THE EXPERIMENTAL CONNECTION

1. A PRELIMINARY: ENERGY UNITS FOR ATOMIC AND SUBATOMIC CONSIDERATIONS

An electron and a proton have exactly the same amount of charge (just the sign of their charges are opposite) and so they gain the same amount of kinetic energy for every volt they move through, albeit in opposite directions. We are talking about a very tiny amount of energy (per particle) here, but it serves as the basic unit in atomic and subatomic considerations. That amount of energy is called an "electron-volt" (eV): the kinetic energy that an electron or proton receives as it moves through 1 volt. In discussions of individual particles, larger units of energy are based on the electron-volt, utilizing the same prefixes that are used in the metric system: "kilo" for one thousand, so that 1,000 eV (10^3 eV) is written 1 keV; "Mega" for one million, so that 1,000,000 eV (10^6 eV) is written 1 MeV; "Giga" for one billion, so that 1,000,000,000 eV (10^9 eV) is written 1 GeV; and "Tera" for one trillion, so that 1,000,000,000,000 eV (10^{12} eV) is written 1 TeV.

These subatomic energy sizes are very small compared to familiar macroscopic energies; a 1 horsepower engine supplies nearly 5 billion TeV of energy each second. To give you some perspective on these energy sizes, let us look at some of the energies we have been discussing qualitatively during this journey. The differences of the energy levels in atoms and molecules, which often yield photons in the visible part of the electromagnetic spectrum, are a few eV. The rest mass energy of an electron is about half (0.511) of an MeV and that of the proton is almost one thousand (938) MeV or 0.938 GeV. Going higher, the mass energy of a W is about 80 GeV, that of a Z is 91 GeV, and that of the *top* quark is about 176 GeV. Our largest accelerators have been able to impart about one TeV (about a thousand times the rest-mass energy of the proton) of kinetic energy to each beam particle and there are plans to go even higher in search of the elusive Higgs particle, and in search of *new physics*.

2. "SCATTERINGS" AND DECAYS

Predominantly, the kinds of experiments that have been performed during the past century in our pursuit of the fundamental structure and interactions of matter fall into two categories: "scatterings" and decays. (The quotes are to remind you that in scattering experiments the outgoing particles are usually *different* than the incoming ones.)

The characteristic common to all of the scattering experiments is that a beam of particles is fired either at a material containing the target particles (called a "fixed-target" experiment) or at another beam. The immediate experimental goal is to find out how many particles of various kinds emerge from the interaction between two particles that "hit" one another, and to determine their respective energies and momenta (including the directions of motion). In fixed-target experiments, most projectiles go right through the target material, just as a stream of clean water would flow apparently unimpeded through a coarse sieve but leave the sieve *wet*. Collisions in beam-on-beam experiments are even more rare because of the diffuse nature of a beam compared to a solid target material. However, for a large number of beam particles, there are often many scattering events to consider. The experimental results consist of the types of initial and final particles and their respective paths, energies, momenta and angular momenta.

For particle theories to be testable in scattering experiments, they must make mathematically precise predictions of the outcomes of a variety of feasible experiments. Furthermore, they must make at least one testable prediction that differs from those of the other extant theories in order to be distinguishable. Because of the inherent quantum nature of the universe, these predictions are usually in the form of probability distributions; consequently, the experimental results of a large number of independent scatterings must be carefully studied and combined. A probabilistic world is a tricky taskmaster!

And what about *decays*? Radioactive substances (nuclei) have been studied by observing the drop of the number of decays per second (e.g. as shown by the changing rate of the clicks of a Geiger Counter) as time goes on. These kinds of decay-studies have been pursued for about ninety years, but they are not the subjects of our discussion here. Instead, we are interested in the direct observation of *individual* events or classes of events, including the properties of *the decay products* of the unstable particles produced.

An unstable *charged* hadron or lepton may last long enough (even tiny fractions of a second will do*) to leave a trail (also called a track) from its point of creation to its decay-end. A *neutral* particle may live long enough to leave a telltale gap between relevant tracks. In both cases, the resulting decay products may then produce a set of trails (by the charged ones) and/or gaps (by the neutral ones) leading to further

*for particles moving at speeds close to the speed of light.

Fig. 19.1. A bubble chamber picture with a sketch, to its right, of the relevant tracks. The reaction is $\pi^- + p \to \Lambda^0 + K^0$, with the subsequent decay in the chamber of each of the neutral particles: $K^0 \to \pi^+ + \pi^-$ and $\Lambda^0 \to p + \pi^-$. Note the tracks that charged particles leave and the telltale gaps between relevant tracks left by the neutral particles. (The bubble chamber picture is courtesy Lawrence Berkeley National Laboratory and the sketch is from "Modern Physics" by K. Krane (1996), reprinted by permission of John Wiley & Sons, Inc.)

trails and/or gaps, etc. You can see both types of paths in the bubble-chamber picture in Fig. 19.1, where an incoming π^- interacts with a proton of the liquid to produce two neutral particles that then decay into charged products. (We shall learn the nature of these trails in the next chapter.)

Because we live in a quantum world, decays follow *probabilistic* laws so that there is a distribution of lifetimes for particles of the *same* type. Consequently, we use the word "lifetime" to mean the average lifetime and "half-life" to mean the time interval during which half of those present initially would decay. We must remember, however, that relativistic time-dilation implies that particles live much longer in *our* frame of reference when they are moving rapidly (with respect to us, of course) than they do in *their* (rest) frame. The lifetimes associated with and listed for unstable particles always refer to times in the respective rest frames of those particles. Because of the time-dilation effect, we have been able to measure tracks for high-speed particles whose lifetimes are as short as 10^{-13} seconds (a tenth of a trillionth of a second).

Most of the particles we have talked about in this book so far have lifetimes that are extremely tiny fractions of a second. Nevertheless, many do leave observable tracks in our detectors because they are usually moving at speeds close to that

of light in our high-energy experiments. Not only would such high-speed particles move observable distances in *those* short times, but the times they actually "live" in our detectors are considerably longer than their (rest-frame) lifetimes for those high speeds — the time dilation effect that we have just mentioned. In fact, particles that leave observable tracks or gaps are considered "stable" in the particle physics world, even though their lifetimes are very short compared to the time intervals that we are accustomed to in our everyday lives; we all know that we mean *stable enough to leave a track or gap* in our experiments.

We can even make high-energy *beams* of the longer-lived "stable" particles, because most of the particles in such a beam live long enough to interact with target particles. Charged pion- and kaon- beams have been routinely used in high-energy experiments even though their lifetimes are only about 1 or 2 percent of a microsecond ($1 - 2 \times 10^{-8}$ seconds). Muons live about 2 microseconds (2×10^{-6} seconds) and copious beams of them have been used in many important experiments.

In contrast, there are particles whose lifetimes are so brief (shorter than 10^{-19} seconds, a tenth of a millionth of a trillionth of a second) that they last *only* long enough to produce their decay-products. They do not leave visible evidence of their existence in any single photograph or other type of recording. However, their existence and their lifetimes can be deduced by studying the momenta and energies of the emerging particles in the data of numerous independent such events. The characteristics of the collection of events that indicate the *blink* of existence of such a particle will be discussed later. Suffice it to say here that these particles are also referred to as "resonances" (of their particle products). Many of the particles that we have met on this journey are, in fact, "seen" in experiments *only* as resonances.

3. THE SCATTERING "CROSS SECTION" — A DIRECT CONNECTION BETWEEN THEORY AND EXPERIMENT

Imagine that you are in front of a wall containing a large number of small balloons distributed randomly over its surface. Now let me *stretch* your imagination to envision those balloons as *microscopic* in size. You are handed a tray full of darts and told to throw them randomly (one at a time, of course) at the wall. The darts have ideally sharp points, so that any hit balloon will definitely *pop*. The number of times you will hear a pop certainly depends on how many darts you throw and the number of balloons on the wall. It also depends on the size of the cross-sectional area ("cross section," for short) that each balloon "presents" to the darts — the smaller the balloons, the less likely that a dart will hit one. Suppose also that *we know how many balloons are on the wall*, and that they are so small that the chances of throwing a dart at a balloon already popped is nil. Then, just by counting the number of pops we hear as you throw the darts randomly at the wall, we can figure

out the size of each balloon. That is because [the number of balloons] *times* [the size of each cross section] tells us how much of the area is covered by balloons. The probability of any one dart hitting a balloon is the fraction of the total wall covered by that area. We expect that same fraction of the darts thrown to give us a pop; the rest is arithmetic.* Thus, *counting* pops indirectly *measures tiny cross-sectional sizes!*

Of course, "scattering" experiments are not dart games but the idea is similar. The size of each balloon gives us its effectiveness as a target, and we need only count events (pops) to know the *effective cross section* of each. We do need to know the concentration[†] of particles in the incident beam (e.g., the number of darts) and the number of target particles (balloons) in the target material (the wall) [or the combined intensities[‡] of the beams, for beam-beam scattering] in order to determine that cross section. All of these properties of the beam(s) and the target (in fixed-target experiments) can be experimentally determined before the scattering experiment is performed.

In many scattering experiments, particles emerge that are different from the incident and target particles, so that the cross section we "measure" (by counting the number of times particular events occur) is not the actual size of anything. However, it is *a direct measure of the effectiveness of the interaction* being studied. And that is what is done in scattering experiments: the number of events of a particular type are counted and used, with information about the beam flux(es) and target density (in fixed-target experiments), to determine the corresponding cross section. These experimentally "measured" cross sections are then compared with the values predicted for them. This implies, of course, that *any theory that is to be tested in these types of experiments must predict the cross sections of observable events.*

The tasks of the experimentalists are to determine the beam flux and target density, to determine what events have occurred, and to correctly select and count the relevant ones. Among these, the event-determination and the correct counting are the most demanding. Many different events may look alike, and it is a major challenge to the experimentalist's skill and knowledge (of both materials and the types of events that can occur) to design and build equipment that can detect and *distinguish* the desired events from all of the others. This winnowing usually involves determining types of particles, energies, momenta (including directions, i.e. angles), etc. (We have seen earlier that even Rutherford's discovery of the nucleus, in 1912, was accomplished by *counting* the number of particles scattered to *large angles.*)

* $\frac{(number\ of\ balloons)\times(cross\ section\ of\ each)}{total\ wall\ area} = \frac{number\ of\ ``pops"}{number\ of\ darts\ thrown}$. The only unknown is the cross section, which can be calculated from this equation.

[†]technically called the "flux," which is the number of particles per unit area in the beam falling on the target in a second.

[‡]technically called the "luminosity" of the collider.

It is also mandatory that we understand the possible *counterfeit* events called "background events" that can evade the technological winnowing, and so will be counted as *legitimate* events. This requires an understanding of the physics going on in and around the experiment and often involves experimental measurements as well.

4. PARTICLE "RESONANCES"

The word "resonance" (and its derivatives) is used in everyday speech in connection with sound, and its meaning has been extended to include more abstract things like concepts and feelings. — "He has a *resonant* voice;" "The sound *resonated* through-out the building;" "Those ideas *resonated* with my own thoughts;" "His dark mood *resonated* with her primal fears." — In physics, the idea has been developed further by considerations of energy. A resonance arises when there is an unusually large transfer of energy from one system to another. This often occurs when a repetitive interaction, like a band marching across a bridge or sound of the right oscillation frequency in an instrument, causes the buildup of large oscillations. Resonances cause instruments or voices to sound full and beautiful, but they sometimes cause bridges to come tumbling down (as the Tacoma Narrows Bridge did in 1940 in the State of Washington).

In our quantum-mechanical world, the energy transfer is often due to the absorption of exactly the correct amount of energy to cause a transition to an excited state of the receiving system. The excited state is even referred to as a *resonant* state or "resonance" for short. We must look at the results of a large number of identical experiments to discern the existence of such resonances and to determine their properties. An appropriate plot showing the *number* of such experiments that produce each value of energy transfer (or a directly related quantity), or a plot of the cross section (for scattering events), shows a peak of the curve corresponding to such a resonance. That peak itself is often referred to as a "resonance" as well.

$$\pi + p \rightarrow \Delta \rightarrow \pi + p \ (\textit{or } n)$$

By the 1950s, physicists had learned how to make beams of the charged pions that were produced in collisions. In 1951, pion-proton scattering experiments had shown a large (resonance) peak in the cross section at a particular energy of the incoming pion. This suggested that an intermediate state had been produced, which subsequently decayed back into a pion and proton (or neutron). The spin, isospin and charge of that *hypothesized* intermediate state were completely determined from the data, and there was no doubt that a resonance had been found. Subsequently, that pion-nucleon resonance was called the delta (Δ).

Before the creation of the quark concept, the Δ was believed to be as elementary as the proton or neutron although too short-lived to leave a track in a detector. In the 1960s the quark model was proposed, in which neither the Δ, the p, nor the

n (nor any other baryon) is elementary but instead each is a bound state of three quarks. And in the 1970s the experimental evidence for the *existence* of quarks came rolling in. In many of those experiments, a large number of other resonances (extremely short-lived particles) were being discovered — both baryons and mesons. We have met some of them; all of the particles in the FLAVOR SU(3) decuplet (10), except for the omega (Ω) which *does* leave a track (as we have already discussed), are resonances. Those nine resonances are understood to be particles that are so short-lived that they *leave no tracks of their existence*.

5. "RESONANCES" AND DECAY WIDTHS

In order to understand *resonances* and their *decay widths*, we will look at the rho (ρ) meson, one of the first spin-1 (vector) mesons to be discovered as a resonance. Its three different charge states, ρ^+, ρ^0, and ρ^-, are three of the particles of a whole [FLAVOR-SU(3)] *octet* of resonance particles. We now know that the lifetime of the ρ is about 4×10^{-24} seconds (four trillionths of a trillionth of a second), which is too short for it to leave any track in our experiments. Let us take a look at an analysis that reveals its existence.

About forty years ago, experiments were being done at Brookhaven National Laboratory (on Long Island, NY) with a pion beam aimed at a large (72 inches across) hydrogen bubble chamber. (The chamber had 500 liters of liquid hydrogen and a building of its own, which also housed computers and the associated electronics to scan particle tracks and store data.) A very large number of various kinds of events resulting from the ensuing pion-proton collisions were studied. The events of interest to us here are those that had *two or more* pions emerging; the incident beam particle was a pion, the target particle was a proton (of the liquid hydrogen), and the emerging particles were a baryon (like a proton or neutron) plus pions.

The distribution of the emerging pions was studied in a very special way. The question was asked: if two emerging pions had resulted from the decay of a very short-lived particle that was produced in the π-p collision, how could we *coax* the data to reveal it? The answer came directly from the conservation of energy and momentum; an algebraic manipulation of the relativistic connection between energy and momentum yielded the *effective mass* of each pair of pions, which is the mass that a particle must have in order to decay into that observed pair.

An effective mass can be calculated for any two pions, even those that have not been produced by a decaying particle. Consequently, we must know how many events are expected, for each value of the effective mass, when no intermediate particle is produced. Then we need to examine the data to see whether there are significantly more events at any value of the effective mass. That *excess* will be the evidence that a very short-lived particle *was* produced in the reaction, which subsequently decayed into those observed particles. The data is usually evaluated by plotting the number of observed events for each value of the effective mass as

Fig. 19.2. A plot of the number of events observed vs. the effective mass of $\pi^+ + \pi^-$ pairs emerging from the interaction $\pi^+ + p \rightarrow p + \pi^+ + \pi^+ + \pi^-$ (from Alff *et al.*, *Phys. Rev. Letters* **9**, 322 (1962), copyright (1962) by the American Physical Society). The dashed curve shows the expected number of events when there is no intermediate particle that decayed into the emerging $\pi^+ + \pi^-$ pairs. Note the excess of events in the peak centered at about 765 MeV. This is a beautiful example of the *resonance* we call the ρ meson. (The jaggedness in the actual data, which has been connected by an appropriate smooth curve here, is as expected for a finite number of events.)

shown in Fig. 19.2. This plot is taken from an original research paper, published in *Physical Review Letters*. Since its discovery in 1962, the ρ meson has been seen as a peak in plots of pion pairs in experiments all around the world, so that its existence is firmly established.

Notice that not all of the excess events in the peak correspond to the same effective mass; instead, there is a spread around the central value. Well, how could that be if these events are all from ρ particles? The answer comes from a ubiquitous property of quantum theory — the energy-time uncertainty relation. A particle with a very short lifetime will necessarily have a large spread of possible values of rest energy and consequently of its mass. When a large number of such particles are examined, the measured mass values will be spread out. The width of that spread, i.e., of the peak in a plot showing the number of events that have produced such particles for a range of values of their masses, reveals the value of the particle's lifetime. The width of the peak in Fig. 19.2 yields 4×10^{-24} seconds as the

approximate lifetime of the ρ. The brevity of that instant tells us, in no uncertain terms, that the ρ decays via the strong interaction.

In conclusion, let me emphasize that the existence of the ρ, as well as its mass and lifetime, can only be surmised by studying *large numbers of events*. No single event shows even a *hint* of its presence.

Chapter 20

PARTICLE ACCELERATORS AND DETECTORS

1. WHY ACCELERATE PARTICLES?

Early in the twentieth century, Rutherford used natural radioactive sources to create the beam in *the first scattering experiments of particle physics*. He discovered the existence of the nucleus using as his beam the α-particles (which are actually Helium nuclei) that emerged from radioactive atoms. At about that time, it was recognized that the Earth was being bombarded by high-energy charged particles from all directions, with a large range of energies, and these have been dubbed "cosmic rays." The cosmic rays were found to have protons, heavier nuclei and also electrons in them, and it was presumed that their origins were galactic rather than local. There is also a stream of particles (mostly protons and electrons) bombarding the Earth from the Sun called the "solar wind," and the name "cosmic rays" is often used more loosely to include *all* of the particles bombarding the Earth's atmosphere. Interactions caused by these particles have been studied since their discovery near the beginning of the twentieth century. Photographic emulsions, similar to the plates used by photographers before the invention of film, were used as the detectors in most of these early studies. The discovery of the muon, and the subsequent discovery of the pion (in the 1940s) as the source of those muons that shower the Earth from the atmosphere, resulted from the study of cosmic rays. However, it soon became apparent that controlled, *mono-energetic* high-energy beams of *one species* of particle (e.g., a proton beam) would allow more detailed nuclear and particle physics investigations.

We expect to learn more about the short-range behavior of the interactions of nature by using higher energy beams, because those beam particles should be able to penetrate deeper into the regions around target particles. Even our intuition tells us that to break an object into its fundamental constituents, we must hit it hard — the harder the better! Furthermore, our understanding of optics tells us that to look at very small things, we must use light of very short wavelengths, and quantum mechanics tells us that *the shorter the wavelength, the higher the*

momentum (hence, energy) of a particle. So "all roads lead to Rome," or in this case to high energy. Furthermore, relativity has taught us that mass is a form of energy, so that higher energy collisions can produce more massive particles — including those predicted by our theories but not yet discovered. The first production of the Ws and the Z gauge particles of the weak interaction required the highest energy available at the time (early 1980s), which was attained at the European Laboratory for Particle Physics called CERN (Conseil Européen pour la Recherche Nucléaire) in Geneva, Switzerland.* Those discoveries convinced all doubters, of the validity of the electroweak theory. The discovery of the *top* quark, in the 1990s, was made at a large accelerator facility called Fermilab,† near Chicago, and *not* at CERN because the CERN collider did not have enough energy to probe the mass region containing the *top*. The search for supersymmetric particles or evidence of other exotic dreams are presumed to require energies thousands of times the rest-mass energy of the proton. And the hunt for the purported "source" of all mass, namely the Higgs particle, is searching higher and higher energy regions. As we do experiments at higher energies, we are exploring unknown regions and may very well be surprised by what we find "up" there!

Cosmic rays certainly have high-energy particles in them, but the corresponding fluxes are often too low to give useful data and their large range of energies makes data analysis problematical. And so, in the 1930s, the first large particle accelerators were designed and built. They utilized the already-understood electromagnetic interactions to accelerate and focus charged particles into well-defined and easily controlled beams.

2. ACCELERATOR TYPES

Cyclic accelerators (synchrotrons)

In the cyclic accelerators used in particle physics, the beam particles go around an oval path again and again (we are talking about millions of times at extremely high speeds), each time receiving a small "kick" until they reach the desired energy. Because the kicks must be synchronized with the motion of the particle bunches around the oval, these accelerators are sometimes called "synchrotrons."

There are two different classes of experiments, usually done at accelerators designed appropriately for them. The oldest type is *the fixed-target experiments*. In

*It is owned and operated jointly by many countries. The member states at this time are Austria, Belgium, the Czech Republic, Denmark, Finland, France, Germany, Greece, Hungary, Italy, Netherlands, Norway, Poland, Portugal, the Slovac Republic, Spain, Sweden, Switzerland, and the United Kingdom. Israel, Japan, the Russian Federation, Turkey, Yugoslavia (status suspended after UN embargo, June 1992), the European Commission, and UNESCO have observer status.
†It was named after Enrico Fermi, a world-class theorist and a premier experimentalist, who created the first quantum field theory of the weak interactions and also produced the first nuclear chain-reaction.

this type, after the beam has reached the desired high energy it is steered towards a fixed target of material. (The target particles are essentially at rest.) Whatever happens next must obey the laws of conservation of energy and of momentum. Consequently, should a single (new) particle emerge from the interaction, it must contain the same energy as the total energy and *the same momentum* as the total momentum of the original pair of interacting particles — a high-speed beam particle and a target particle. I have emphasized "the same momentum" because this tells us that not all of the incident energy is available as rest-mass energy for a new particle. The created particle must be *moving* with the same momentum that the beam particle brought into the interaction, so that some of the incident energy becomes the kinetic energy of the particle produced. This is a severe limitation on the fraction of the initial energy available as rest-mass energy, consequently severely limiting the mass of the particle that can be created. Clearly, a search for *high* mass particles with fixed-target accelerators is not at all efficient; for an enormous amount of (initial) energy must be supplied, most of which merely conserves momentum and so is not available as mass energy.

Now let us consider a "collider." In this type of accelerator, there are *two* beams. In the regions where these beams cross, we will have some interactions between particles of one beam and those of the other. In such an interaction, each particle has the same amount of momentum but they are traveling *towards* one another just before the interaction. Their momenta are equal but *oppositely directed*, so that the total momentum of the pair is zero. This means that when a single particle is produced by the interaction, it will be produced at rest, i.e. also with zero momentum. — It's just conservation of momentum! So, *all* of the initial energy is transformed into the rest-mass energy of the particle created. Yes, these colliders can create particles of impressively large masses — but beam-on-beam collisions? Aren't they very rare? Indeed, they are; we are much much more likely to have a "collision" when we aim our beam at a solid target. But that is the tradeoff. For studies that require a very large number of events (to get great detail), we need a fixed-target machine. To search for and produce very heavy particles, a collider wins, hands down! And more and more intense beams *are* being produced at colliders to raise the collision rates to acceptable levels for some detailed studies. In order to enhance the intensities of the colliding beams, they are made as narrow as is technologically feasible. The beam diameters attained are of the order of millimeters — the thickness of an ordinary piece of cardboard.

Beams of stable particles can move around and around in colliders for days; sometimes their intensities are increased as more and more particles are added. This storage aspect led to colliders also being referred to as "storage rings." The storage requirement has restricted the primary beams of colliders to truly stable particles like protons, antiprotons, electrons and positrons. There is, however, current talk about the feasibility of muon colliders. Although the half-life of a muon is only a couple of microseconds (2×10^{-6} sec.) that may be long enough for muons of high

energy because of the time dilation effect: high speed particles live a much longer time in the lab frame than their (rest-frame) lifetimes would imply. Furthermore, with an intense beam there may be enough particles left after several half-lives to produce a measurable number of events.

There once was a dream to build a giant collider that would be able to provide about 40 TeV of rest-mass energy. It was called the Superconducting SuperCollider (SSC). That collider would have provided definitive answers to our questions about the existence of the Higgs particle, as well as sundry other theoretical conjectures. Its construction was already well underway in the early 1990s when it was canceled by the US Congress. The highest energy collider on the horizon is being built at CERN in Geneva. It is called the "Large Hadron Collider" or LHC.

One drawback of cyclic accelerators has actually been put to good technological use. In a synchrotron, there are actually two kinds of acceleration taking place. There are those kicks that increase the energies of the particles each time they traverse the course, but there is also the acceleration required to keep the particles moving in the oval path. Even a particle moving at a *constant* speed around a circle is undergoing acceleration since its momentum is changing *direction* — a particle that is not accelerated at all moves in a straight line (recall the law of inertia). Beam particles are kept on the oval path by magnets that exert the appropriate forces. The significance of this to our energy considerations results from the fact that *accelerated* charged particles radiate electromagnetic energy. Consequently, a lot of the energy supplied to the particles by the kicks is radiated away because of the acceleration keeping them on the oval path. This *synchrotron radiation* has been put to good use however, especially in medical applications. Unfortunately, it does ruin the efficiency of cyclic accelerators to produce very massive particles (without running up an impossible utility bill).

Linear accelerators (linacs) and linear colliders

For electrons, synchrotron energy losses are so severe that a very large *linear* accelerator was built to produce high-energy beams. It is the Stanford Linear Accelerator, located at the Stanford Linear Accelerator Center (*SLAC*) in California.* The limitations imposed by synchrotron radiation on the ability to produce high-energy beams, in fact, preclude a cyclic design for any future high-energy accelerator. So, the next accelerator will (probably) be a *linear* accelerator that is referred to, for the time being, as the Next Linear Collider (*NLC*).

Accelerator dimensions

The accelerators built in the 1920s and 1930s were small enough to be contained in a room; however, as the desired energy of the accelerated particles grew, the size of the

*A collider called the Stanford Linear Collider (SLC) was later added utilizing that linear accelerator.

Fig. 20.1. Aerial view of CERN (courtesy CERN) in Geneva and environs. The large white (thin) oval shows the location of a tunnel in which the largest accelerator ring is located. It is 27 kilometers (about 17 miles) in circumference. The small white crosses were inserted to indicate the border between Switzerland and France, (mostly) below them is Switzerland, above is France.

corresponding accelerators reached the size of miles (or kilometers, if you are so inclined). The SSC that was being built in Texas was to be 56 miles in circumference. In Figures 20.1, 20.2, and 20.3, we see aerial views of the accelerators we have discussed above: CERN (near Geneva, Switzerland), Fermilab (near Chicago, Illinois — USA), and SLAC (near Stanford, California — USA), respectively.

3. THE BASICS OF PARTICLE DETECTION

Detectors are employed to ascertain the properties of the particles that emerge from the interaction being explored, and, in scattering experiments, of those in the initial beam(s) as well. In order for a device to detect a particle, it must contain material that interacts with that particle in such a way that the particle's properties can be easily discerned.

The macroscopic behavior of charged particles, moving through a medium in the presence of electromagnetic fields, is well described by the same classical

Fig. 20.2. Aerial view of Fermilab (Courtesy of Fermilab Visual Media Services), with the Main Injector accelerator in the foreground, and final high-energy accelerator (called "Tevatron") in the back. Fermilab sits on a 6800-acre site. The rings are about 4 miles in circumference, and are easily seen from planes taking off or landing at O'Hare International Airport.

electromagnetic theory that inspired the development of radio, television, radar, microwave ovens, electrical power and the electronics of the first half of the twentieth century. Consequently, electromagnetic interactions form the basis for the design and operation of many detectors. Furthermore, once *any* effect has been studied and understood, it can also be used in the detection and study of new effects; this has, indeed, occurred in some of the explorations in particle physics. In general, we can say that detectors employ previously studied effects in their design and use.

Time measurements, along with various geometrical arrangements of the basic devices that we shall discuss, are used in innovative ways that are appropriate to the particular experiment being performed.

Fig. 20.3. Aerial view of SLAC (courtesy Stanford Linear Accelerator Center, funded by the US Department of Energy). The white rod-like shape extending upward towards the right from the center of the picture is the two mile long accelerator tube (actually running under an expressway).

Particle trails

A charged particle moving through any medium (a gas, a liquid, or a solid) exerts electromagnetic forces on the electrons in the atoms and/or molecules of that medium. These forces are so large that they rip electrons out of those that are along their paths, leaving them with a net positive charge. They are then referred to as ions and this process is called "ionization." The path of the particle is thus "remembered" for a while by the medium in the form of ion and electron trails. There are several ways that the path can be recorded before or as the ions or freed electrons disperse and these methods are incorporated in several different kinds of detectors.

In many of these devices, a very large magnet is employed in order to determine the momentum of the charged particles entering and/or emerging from an interaction. (No need to worry about such quantum limitations as the uncertainty principle here, for they are completely undetectable in these macroscopic detectors.) The magnet is set up so that the particle paths are all inside the region that contains the magnetic field. A magnetic field causes moving charged particles to spiral around its field lines, the faster a particle is moving perpendicular to those lines the larger is the radius of its spiral path; furthermore, positively-charged particles spiral one way, negatively-charged particles the other. For a high-energy particle, the radius of the spiral is so large that the path we see is only a small part of it, looking like a curved line. A measurement of the curvature of that path (plus directional information) tells us the particle's momentum, and its orientation tells us the sign of the charge.

There are many different variations of some basic schemes that are employed in the design of detectors. Let us now peruse the more prominent examples of these schemes.

Meet the detectors

We will start with the venerable "bubble chamber." Many of the discoveries made in the early years of development and verification of the Standard Model were made in bubble chambers. These chambers contain a very pure liquid that is heated above its boiling point; because of its purity, it does not have centers about which bubbles can form until a charged particle moves through, ionizing some of the molecules it passes. The ions produced serve as centers for bubble formation, yielding a bubble trail of the particle's path. Stereographic photography allows the reconstruction of the three-dimensional paths of the charged particles and the subsequent determinations of their momenta (from the curvature of their paths). Bubble chambers were used in many important high-energy experiments from the 1950s to the 1980s.

More modern devices, known as "proportional counters", "drift chambers" and "time-projection chambers" are used in innovative combinations in today's detectors. They are all small, basically very thin chambers filled with gases, containing one or more high-voltage wires. When charged particles pass through them ionizing the gas, the released ions and/or electrons are attracted by the wires. Electronic readouts produce the desired information detailing which of the counters the particles went through. By combining the readouts from many small detectors, the particle paths can be determined.

There are also devices, somewhat related to the proportional counters, that produce *visible* sparks right in the chamber instead, showing the particle paths directly; they are "spark chambers" and "streamer chambers." Some devices use solid materials, both inorganic and organic, that emit visible light as a charged particle moves through. (The moving charge excites the atoms in its path; each

excited atom then "relaxes" via a transition from the excited state back to the ground state with the concurrent emission of a photon in the visible range.) These substances are used to construct "scintillation counters" of various shapes. Another type of solid-state device often used as a path detector is the "silicon strip detector." These are made of the same kinds of chips used in the ubiquitous solid-state devices that have been driving the technological revolution of the past forty years. In these chips the freed electrons produce small, detectable currents. The sequence of chips hit, along with the location that has been struck on each such chip, shows the path of the charged particle that went through.

Now we come to a device that employs a completely different electromagnetic effect, called a Čerenkov counter. It employs the Čerenkov effect, named after the Soviet physicist who discovered it in 1934. Indeed, the equations of electrodynamics show that when a charged particle moves at high enough speed through a medium, it radiates a cone of light in its wake; the opening angle of the cone tells us the velocity of the particle. This occurs because, in a medium, the velocity of a light wave is less than c (the speed of light in a vacuum) due to interactions with the molecules of the medium; in water for example, it is $3/4 c$. Particles moving faster than that light-wave speed produce that telltale cone of light in their wake, which is easily detected by surrounding phototubes. Thus Čerenkov counters can give us the velocity of a high-energy charged particle. If we have independently determined the *momentum*, we can then calculate the mass of the particle. Clever geometrical arrangements of these counters can be used to obtain images of the particle tracks as well.

Energy determination

In bubble chambers, (nuclear) photographic emulsions, or solid materials, the energy of some particles can be determined from a measurement of the distance that they move before stopping. More sophisticated techniques are used for very-high-energy particles and the corresponding devices are called "calorimeters." The word "calorimeter" originally was used for devices that measured heat, but it has been generalized by particle physicists to refer to devices that measure the total energy of an absorbed particle. A typical calorimeter looks like a multi-layered sandwich — bread, meat slice, bread, meat slice, bread, etc. The "meat" slices in this sandwich are only a few millimeters thickness of lead, or another heavy metal. But a few millimeters (the thickness of a piece of cardboard) is quite a large distance on the microscopic scale. In fact, much of the entering particle's energy is lost in each slice via ionization; after going through several metal layers, the particle has lost virtually all of its kinetic energy. The electrons knocked free in the metal are so energetic that they themselves ionize more atoms and the newly freed electrons can do the same — and so on, and so on, and so on. The electrons can also emit photons during their encounters, and some of the photons can create electron-positron pairs

(moving at high energies) which can also continue the ionization process as well as produce more photons, and on, and on. — ZZAAAP — An electromagnetic shower is produced in the metal layer. The "bread" layers of the calorimeter are about a centimeter (half an inch) thickness of *scintillator* material that the showers activate to emit light. Phototubes then measure the amount of energy thus emitted. The geometric arrangement of metal and scintillator can be varied, so that besides the (sandwich) *tower calorimeters* we have just described, there are specialized designs such as *spaghetti calorimeters* — a thick metal "post" with long strands of scintillator embedded lengthwise, the ends of the rod looking just like the open end of a box of spaghetti. So, sometimes it's a tower sandwich and sometimes it's spaghetti — *bon appetit* to our experimentalists.*

Mass and type

Combining any two of the energy, momentum and velocity measurements allows us to calculate the mass of each particle, so that we can determine what kind of particle it was — an electron or a π^-, or etc.; a positron or a π^+, or etc. In some experiments only the momentum value or only the energy (and direction of motion) is measured, so that other information is sometimes needed, such as what particle it *should* be as determined from other experimental or theoretical information, or the density of the bubbles or sparks in its ion trail, etc.

Electronics galore

And let us not forget the electronics! Masters of the art of electronics, our explorers adapt existing equipment and create new designs, as needed, to accomplish their goals. Several of their innovations have been adopted and adapted by the consumer electronics industry. As we shall see soon, high-speed electronics, including computers, is essential in order to be able to find the "wheat" hidden in the mountains of "chaff" produced in high-energy scattering experiments.

Computer software

In the course of designing a large-scale high-energy experiment, computer software must be created by the researchers. The randomness of the locations and momenta of the particles in the beam(s) and those of the target particles are built into some of the computer programs in order to predict the final distributions one might expect. These kinds of programs are called "Monte Carlo programs" — named after that famous gambling mecca. Physicists also write software to serve as the researchers' surrogates in making rapid decisions as to which events should be fully recorded

*There are several other ways to detect electromagnetic showers — proportional counters and streamer chambers (mentioned above), as well as several others, but they are not as delectable to describe.

and then in recording the details of those events — a multi-talented lot, these experimentalists!

4. LARGE HYBRID DETECTORS

Most particle experiments are done at accelerator facilities that are as large as villages, or are set up at similar sized facilities designed to detect neutrinos. In either case very large detectors are used, which are generally as big as multistory buildings.

Just about all of the detectors at accelerators are complex, hybrid creatures containing many different layers of detector modules of various sorts laid out in strategic geometrical arrangements. Each structure is permanently located at one particular experimental site. They are such enormous, complex facilities that they require an army of physicists, engineers and technicians to design, then build, as well as to operate them during experiments. There are about twenty such enormous multi-layered detector facilities at the large accelerators around the world. Their names are often acronyms, like those of the accelerators themselves, as is a common practice nowadays (like "FANNY MAE", "FBI", "CIA", etc.) Each has its own particular structure, suitable to its purposes and the accelerator at which it operates.

An example of a detector at an accelerator — the CDF at Fermilab

The CDF is a huge detector located at Fermilab (CDF is an acronym standing for "Collider Detector at Fermilab"). It weighs 5,000 tons and is made up of an enormous magnet (that curves the paths of charged particles to determine their momenta) and many, many layers of detection. It is about as tall as a three-story building. CDF is composed of a very large, elaborate tracking system along with complex calorimetry and muon systems each containing an enormous number of individual detectors — 130,000 electronic channels in all. Over 50 institutions from about a dozen countries are involved in the experiments. Figure 20.4 shows the detector in its maintenance hall with its calorimeter arches pushed to the sides so that it can be serviced. To appreciate its size, note the person standing on the floor towards the lower left-hand corner of the picture; above, there is another person standing on the right side of the central circular region. This detector was designed and built with such meticulous precision that its thousands of very sensitive and accurate small detector units operate in an astonishingly coordinated manner. Its nearly flawless performance is a monumental tribute to those who designed it, those who built it, and those who maintain it. Among its many different parts can be found time-projection chambers, multiwire proportional chambers, drift chambers and many fine-grained calorimeters. It is a wonder to behold!

Fig. 20.4. The CDF (Collider Detector at Fermilab), with its calorimeter arches pulled to the sides for maintenance (courtesy Fermilab Visual Media Services). When it is placed in the accelerator, the beam pipe goes through the center of the central circular region.

A European example of a detector at an accelerator — DELPHI at CERN

The Large Electron Positron collider (LEP) at CERN is no longer operational as it is being replaced by the higher-energy Large Hadron Collider (LHC), which will be placed in the same tunnel. However, let us take a look at one of its detectors to get a sense of the scale and complexity involved. The beam pipe of LEP at CERN had four regions where the electron and positron beams collided head-on. DELPHI (DEtector with Lepton, Photon and Hadron Identification) was one of four huge detectors, each located at one of those places in the LEP tunnel. This detector contained a huge superconducting electromagnet (in a tubelike shape) that curved the paths of the charged particles being detected, so that their momenta were easily determined. The paths were measured by many layers of detectors, so that the directions of the momenta were accurately determined. It employed Čerenkov devices to measure velocities, which yielded good particle identification when combined with momentum information. DELPHI also contained many different kinds of chambers forming its tracking detectors as well as calorimeters, and scintillation counters for particle identification. Most information supplied by the detector was

Fig. 20.5. DELPHI (DEtector with Lepton, Photon and Hadron Identification) before installation (courtesy CERN). When in place, this detector surrounded one of the four collision sites of the electron and positron beams; the LEP beam pipe ran right through the center of this cylindrical detector.

in three-dimensional form, read out via about 200,000 tiny electronic channels by 72 small microprocessors (the tiny "brains" of computers). The information was finally sent to a central set of larger computers to be prepared for storage. About one million bits of information were stored on magnetic tape for a typical particle event studied. This 3,500-ton detector took 7 years to design and build. About 700 physicists, engineers, and technicians from 56 universities and institutes in 22 countries were involved in its operation. Teams from those laboratories have spent about a decade collecting and analyzing data obtained by this masterpiece of precision; the analysis is still ongoing. Let's visit DELPHI by looking at Fig. 20.5.

State-of the-art neutrino detectors — the Sudbury Neutrino Observatory & Super-Kamiokande

There are several neutrino detectors in operation (around the world) studying the neutrino flux from the sun, some for many years. Since neutrinos do not have strong or electromagnetic interactions, it requires an enormous amount of detector material to obtain enough interactions to reach definitive conclusions. Furthermore, many

interactions of cosmic rays with the materials in and around the detector can mimic the neutrino interactions being studied. Consequently the detectors are located either deep in a mine, under a mountain, or in a tunnel to shield them from those cosmic rays. They have all found a paucity of electron neutrinos coming from the sun compared to the prediction of the well-accepted (observationally supported) theory of the behavior of stars. We have already seen in Chapter 17 Section 1, that we can account for their results by considering neutrino oscillations. As we mentioned there, the Sudbury Neutrino Observatory (SNO) and Super-Kamiokande have both found evidence of the flavor-oscillation of neutrinos (with their results complementing one another), with the implication that neutrinos are not massless after all. Super-Kamiokande utilizes the Čerenkov effect to detect high-speed charged particles in its enormous tank of pure water. It is 40 meters (about 131 feet) in diameter and 40 meters tall. Thousands of sensitive phototubes line the inside walls of the vessel to detect the Čerenkov light.* The most recently built neutrino detector, SNO, is located in a mine in Sudbury, Ontario, Canada. It contains 1000 metric tonnes of heavy water and an array of 9600 photomultiplier tubes. Including the outside supporting structure, it is the size of a ten-story building.

5. IMPORTANT EXPERIMENTAL CONCERNS

Signatures, triggers and vetoes

When an experiment is in progress, many "illions" of events occur per second within the large hybrid detector on the scene. The experimenters are usually interested in only about 1 out of every 100,000 of those, and there is not even enough *room* to store detailed information from all of the "dud" events. Consequently, only some fast-reacting parts of the detector are on all the time, to feed information to the decision-making computers. These computers have software programs designed by physicists to recognize certain special features (called "signatures") of the kinds of events being studied and then, if they are detected, to turn the rest of the detector facility on. Included in these programs are instructions to *veto* the turn-on and recording, if certain unacceptable features are detected. The quick-response parts of a detector plus the software containing the signatures of the desired events, etc., are collectively referred to as the "event trigger."

Let me concoct a hypothetical, simplified scenario, just to make this more concrete. Consider a search for events in which a muon is created. There are an enormous number of muons bathing the Earth from the atmosphere all the time, and surely some of them penetrate the detector.† Muons are such highly-penetrating

*Unfortunately, several thousand of them imploded (in a chain reaction) on Nov. 12, 2001. Rebuilding is anticipated.

†They result from protons in the solar wind or cosmic rays that collide with the molecules of our atmosphere and produce pions, which then decay to muons, etc.

particles, that in the early days (the 1960s), steel from the hulls of surplus battleships was used to stop all particles *but* muons from reaching the outermost detectors, consequently called "the muon detectors." Part of the trigger program would search for the ionization trail of a charged particle and an accompanying hit in one of the muon detectors. One "veto" might be an initial signal in a scintillation counter just above the main chamber indicating that the detected muon probably entered the chamber from *above*, so it wasn't created in the interaction after all.

There is a possible danger lurking in all of this automated filtering of data — Could the triggers be keeping new, as yet undiscovered, physical interactions from our view? We hope that we are not losing new kinds of "wheat" in the discarded "chaff." Those writing the software are presumably being as careful as they can to avoid "throwing out the baby with the bath water," but there may be some surprises in store for those who try the *unusual*.

Backgrounds

Extraneous muons are not the only sources of contamination of the data. No matter what signatures of an interaction we search for in our experiments, there are always other events — some due to cosmic rays, some just accidental look-alikes — that have the same signature. These *impostors* are collectively referred to as "the background." Sometimes, a good understanding of the detailed distributions (in angle, momentum, or etc.) of various types of events allows us to "cut" out portions of the ranges of the measured quantities so that many of those background events are deleted from the analysis. The remaining data is then *richer* in the events desired. Ultimately, we must know approximately how many background events to expect, so that we can correctly interpret the data. We have already talked about background considerations in our discussion of the discovery of the ρ meson; it is apparent that without an understanding of the background (remember the dashed curve), the data would have been useless.

Counting rates and cross sections

We are all mortal and our research careers, especially nowadays, require frequent published results. A very cleverly designed trigger for a detailed signature might indeed omit most background events, but we would hope that it doesn't suffer from another "disease" — namely, a *low counting rate*. Consider a setup that is expected to produce about one event per year; would you undertake such an experiment? It might take 30 years before you would have enough data to draw any conclusions at all. So, counting rates *count*! The counting rate, of course, depends on the intensity of the beam(s) as well as the cross section (likelihood of occurring) of the events sought. As we go up in energy, many new particles can be created, so that the number of *types* of events grows rapidly; consequently, the cross section for any one

type usually drops. Higher energy, then, may not be the right answer for certain studies; more intense beams might, instead, fit the bill.

Detector efficiencies

Real detectors, made of real materials, are not anywhere near 100% efficient. In fact, their efficiencies at detecting particles (or energies, etc.) vary with the energies, directions, and other properties of the particles. It is the responsibility of the experimenters to study their detectors responses so that they intimately understand the limitations of each detector they design and use. This is a responsibility that is thoroughly and admirably fulfilled by our elementary particle experimentalists, so that their analyses of the data of their experiments are certainly reliable. Since almost all of the high-energy experiments involve many experimenters, there is certainly more than one physicist involved in the analysis of detector efficiencies, so that everything is *at least* double-checked.

All of the above considerations must be taken into account in deciding the feasibility of a proposed experiment at a particular facility and the designs of new facilities.

FINALE

SYMMETRY blink GRAVITY

Before the shiny worms did writhe and wriggle in the mud,
before that mud was muddy and before the putrid slime,
before the seas receded and the land stretched out in light,
before the Earth was formed 'round searing sun,
before ignition of the sun shot forth the blinding light,
before the elements were brewed in hot primordial stars,
way back when there was no "before," and time had just been born,
And likewise space was bursting into being,
'twas then that *dot*, the universe, had symmetry so grand,
and brought forth from its energy a joyous multitude
of particles.

And many types of particles cavorted in that space,
And there were quarks and there were leptons, zooming here and there,
And they were always interacting with the spin-one ones.
(Some say they were not particles at all,
but strings minuter than minute vibrating on their own,
that look like particles from far, and *far* we'll always be.)
Some spin-one ones were gluons, and the photons were there too
and *W*'s and *B*'s* were all around,
and also others with no names, who've hidden from our view.
But all obeyed the symmetry as space flung out and out,
and cooled.

It was a symmetry so grand that reached to every spot,
and treated all those quarks and leptons equally to joy
of interaction with the spin-one ones pervading space.
With perfect rhythm, they all gaily swayed.
No mass was there, as spin-one ones would nakedly cavort
among themselves, and with the other particles they danced.
The quarks and leptons flew so wild for they were massless points
and reveled in the spin-one ones' delights.
And of disaster, devastation soon to burst the scene
no hint right then, 'midst all the fun, though there's no doubt that 'twas
surely predestined!

continued over

*B is the name given to the gauge particle of the U(1) of hypercharge.

They reveled in their freedom — yes, unfettered, unrestrained —
but all the while the universe unfurled in fur(i)ous haste.
And it did cool and cool and cool to Higgses' great delight,
for they, in turn, could *thicken* all of space.
And that is how they plucked brute "mass" right out of space itself,
thus breaking down the symmetry as "mass" came into being
and splattered on the spin-one ones with horrible, hot spite.
At first small symmetries remained unbowed.
But then that 'lectroweak one was most cru-e-lly untwined
by Higgses flinging "mass" about, like mudballs that then smashed
and spin-one ones defiled!

The bits of shattered symmetry that still reigned on stood tall,
their pride shined faintly 'mid the shards that lay there all around.
Electromagnetism did survive along with Strong
the brutal chaos of the sudden fall.
Their spin-one ones, the photon and strong-colored gluons pure,
remained untouched, unsullied by debilitative "mass."
The quarks were all consoled that color-charge was still conserved,
ignoring warnings of those gluons' might.
And so they gathered in small groups to celebrate that strength,
extolling with conceit and such proud haughty arrogance
the power of "color."

But then those quarks could not resist the gluons rav'nous grasp.
And so their liberty was lost, despite their plaintive cries,
their freedom gone as they were all entombed in gluon bags,
no more to roam, no more to dance about.
But gluons' powerful crushing grips turned on themselves as well —
no *single* gluon could exist though two would clench entwined
and hide their color from the world as glueballs they did form.
Thus color-charge of quarks' and gluons' pride
in glueballs and quark-hadron bags are hidden from our view.
Most hadrons then decayed away, just protons 'n neutrons left,
later to be cooked together.

continued over

But leptons are all colorless, immune to gluons' songs,
so they could still fly here and there with full impunity.
And free most are today as well, except for those with charge.
For photons' songs enchant all charge e'en now,
with their bewitching, 'xotic tunes that do both pull and push
electrons, protons and the rest, beguiled by wistful strains.
Some charge was minus, some was plus frustrating each's effect.
Most 'lectrons were enticed by nuclei
and that embrace did form the atoms 'n molecules, uncharged.
Thus charge was by and large erased, though easily brought forth,
but *mostly* hidden.

Well when that blink* of time was gone, grand symmetry undone,
just colorless and neutral dust and splinters then remained.
And lo the weakest force of all stood out above the rest;
'twas gravity that then held sway o'er all.
Its subtle secret 's that it never can be neutralized.
The "power" is that it just pulls, no pushes canceling.
It can't be hidden, can't be quenched and so it reigns supreme.
And all the teeming galaxies of searing stars that be,
and crunching great black holes of song and verse,
are glor(i)ous children of its loins as now it does dictate
the order of our universe.

*The time for the grand symmetry to be broken and hadrons to be formed is thought to have been extremely short, but the time for neutral atoms to dominate is estimated to be roughly 400,000 years. However, the universe is roughly 14 billion years old (14×10^9), so that the time it took was about 3 one-hundred-thousandths (3×10^{-5} or .00003) of the age of the universe, about the same ratio as the blink of an eye to an entire day — with poetic license, it is surely a blink of time.

EPILOGUE

It was my good fortune to have been enabled, by the generous financial support afforded me as a student, to learn the profoundly beautiful concepts of modern physics and their experimental validations. I had the privilege of attending classes and seminars presented by some of the most creative scientists of our time. In the years that followed, I was also fortunate to be surrounded by talented colleagues and research collaborators, who have broadened and deepened my understanding. And now I have enjoyed taking you, dear reader, on this fascinating journey exploring our understanding of the underlying workings of nature. I hope that *you've* enjoyed our excursion as well.

There *are* those who charge that modern science is merely the *creation* of ordinary human beings who, above all, have egos and reputations as their guiding principles. Well, of course they're right, as far as that characterization goes. All of the physicists I know or have heard about are or were human beings. We all have preconceptions, ambitions, emotions and all of the other human attributes that are so often maligned by critics of modern science. There is definitely a sociological component to what types of experiments, or theoretical notions, are "in" at any particular time and place. Rumors about "hot" new experimental results (or even about unreported theoretical breakthroughs) sometimes do, indeed, arouse anxieties and even affect behaviors. However, the conclusions drawn by those critics concerning the validity of modern science are dead wrong! We have no unassailable writ or dogma! In fact, there is nothing that whets the appetite of a research scientist *more* than the prospect of finding something completely *new* and *revolutionary*. Furthermore, we all have *the discipline of the scientific method*, and *the verdicts of nature* in our experiments, keeping our professional work from corruption. *Logic, consistency, and "hard" experimental facts do not bend to our whims or failings.* Instead, modern science stands as one of the crowning intellectual achievements of our species — *the universe questioning itself*. Yes, we are without doubt a remarkable *part* of the universe that we are exploring, and our achievements in scientific understanding are truly *amazing*.

GLOSSARY

absolutely conserved conserved by *all* of the interactions of nature.

accelerator a device that increases the velocity of the particles that enter, subsequently emitting a beam of high velocity, high energy particles.

angular momentum a quantity that is a measure of the rotational motion of objects.

anti- a prefix that refers to another particle of exactly the same type, but with opposite sign charge and opposite values of any other chargelike properties. Also used to directly refer to opposite color properties, such as anti-red etc.

associated production the phenomenon of kaon production in all low energy experiments that produce heavy baryons (cousins of the nucleon).

atom smallest (neutral) particle of any element; composed of a small, heavy nucleus surrounded by electrons.

B baryon number.

background events detected in an experiment that look like those being studied but whose sources are not the types of interactions under investigation.

baryon number (B) the total number of baryons *minus* the total number of antibaryons. In terms of quarks, it is *one third of* ([the total number of quarks] *minus* [the total number of anti-quarks]).

baryons subatomic, strongly interacting particles like the proton and neutron, that have odd half-integer spin (consequently fermions); composed of three quarks.

beam pipe in an accelerator; a tube, usually looking like a long water pipe, inside of which the much narrower beam travels.

beauty a flavor of quark, also known as bottom.

beta (β) particle an electron (β^-) or positron (β^+).

beta decay the changing of a nucleus of one type (element) into the nucleus of another, with the concomitant emission of an electron or positron and an antineutrino or neutrino, respectively.

boson a particle whose spin is an integer (times \hbar).

Big Bang the explosive event presumed to have occurred at the beginning of the existence of our universe (which started as a point).

bottom a flavor of quark, also known as *beauty*.

bubble chamber a detector containing a (pure) liquid so hot that bubbles form around the paths of charged particles that have traversed it.

C the symbol representing the operation called charge conjugation.

calorimeter a device that measures the total kinetic energy that a particle had when it entered the device. (This word was originally used for devices that measure heat flow.)

cascade (Ξ) an isodoublet (two particles) of baryons, in the same SU(3) octet as the nucleon.

CDF acronym for Collider Detector at Fermilab. A huge, complex detector operating at the Fermilab accelerator facility.

CERN (Conseil Européen pour la Recherche Nucléaire) The European Laboratory for Particle Physics; a very large, complex accelerator facility located in Geneva, Switzerland. It is owned and operated jointly by many countries: member states are Austria, Belgium, the Czech Republic, Denmark, Finland, France, Germany, Greece, Hungary, Italy, Netherlands, Norway, Poland, Portugal, the Slovac Republic, Spain, Sweden, Switzerland, and the United Kingdom. Israel, Japan, the Russian Federation, Turkey, Yugoslavia (status suspended after UN embargo, June 1992), the European Commission, and UNESCO have observer status.

charge a property of particles that causes them to experience electromagnetic interactions (mediated by photons).

charge conjugation the (conceptual or mathematical) operation that replaces each particle by its antiparticle.

chargelike for a conserved number: being additive for a collection of particles (just as electric charge is).

charm a flavor of quark.

Čerenkov effect an electromagnetic effect, whereby light is (photons are) emitted as a charged particle moves through a medium faster than light would in that medium (but always less than the vacuum-speed of light). Detectors that employ this effect are called Čerenkov counters or detectors.

clockwise a rotation in the same direction as the motion of the hands of a clock, or located in the same positions as the orientation of the numbers on a clock.

collider a particle accelerator that accelerates two beams and causes them to collide with one another.

color the property of particles that causes their strong interaction (with gluons). It is not related to visual color in any way, but is certainly a *colorful* way to describe this property.

color neutral colorless, white, or a color singlet.

color singlet a particle that is not changed (or mixed with others) by color-SU(3) symmetry operations. See *colorless* and *white*.

COLOR SU(3) or **$SU_c(3)$** the gauge theory of the strong interaction. Gluons are the gauge particles. See *gauge bosons*.

colorless The designation of a particle or set of particles that are unchanged by COLOR-SU(3) transformations, i.e. a color singlet. Also referred to as *white*.

commutation the act of interchanging the order in which two operations are performed, so that the second one is done first.

commutative group a group of operations, all of which commute with one another.

commute two operations commute if the results obtained by performing them successively in different orders are the same, i.e., applying operation (1) followed by operation (2) gives the same result as (2) followed by (1).

confinement a property of an interaction that prevents the appearance of the "charges" (colors in quantum chromodynamics). It makes it impossible to break bound particles from their bondage. Quantum chromodynamics possesses this property, so that all particles that are actually observed are colorless.

conservation law a law having the property that the total amount of the quantity in question remains fixed despite interactions. We express this by saying that the quantity obeys a *conservation law*.

conserved quantities quantities that obey a conservation law.

constructive interference waves adding together to produce a wave larger than either one.

continuous a property of some infinite sets such that between any two members there is always another member of the set; consequently, there are an infinite number of members between any two. The points on any line form a continuous set, and so do the numbers running from 0 to 1. (Between .9999 and .9998, we find an infinite number of numbers, e.g., .99981 , .99982 , .9998001, .99980001, etc.)

counter-clockwise the direction opposite to clockwise.

correspondence principle in quantum mechanics: the principle stating that results obtained in quantum mechanical calculations should become more and more like the results obtained in classical physics as we raise the values of the relevant quantum numbers. More generally, a requirement imposed on any theory that its predictions agree with those of an already successful theory in the range of validity the latter.

cosmic rays high-speed particles bombarding the earth's atmosphere from outside our solar system, sometimes used loosely to include all those bombarding the atmosphere. See *solar wind*.

CP the conceptual (and mathematical) operation that simultaneously replaces each particle by its antiparticle and considers the mirror (or inversion) image of the interaction or event under consideration.

CPT the conceptual (and mathematical) operation that simultaneously replaces each particle by its antiparticle, considers the mirror (or inversion) image of the interaction or event under consideration, with reversed time order.

cross section the effective-target-size (not necessarily connected to physical size) for a given event (or class of events) that represents how effective the interaction is in producing it (or them).

cyclic accelerator here, a particle accelerator in which the beam particles traverse the same oval path again and again receiving small "kicks" during each traversal.

De Broglie wavelength the wavelength of the wavefunction that describes a quantum particle.

decuplet here, a grouping of ten particles that form a multiplet of SU(3).

DELPHI acronym for DEtector with Lepton, Photon, and Hadron Identification; a massive complex detector that was located at CERN in Geneva and used when the LEP accelerator operated.

delta (Δ) a (spin-3/2) baryon that comes in four different charge states (Δ^{++}, Δ^{+}, Δ^{0}, and Δ^{-}). It is very short-lived and shows up only as a resonance. It is the isospin-3/2 member of the FLAVOR SU(3) decuplet.

destructive interference waves adding together to produce a wave smaller than either one.

deterministic philosophically: an adjective describing a world in which the present completely (precisely) determines the future.

diffraction the spreading of a wave after going through an opening or passing a barrier.

discrete isolated; not a continuum. A *discrete* group is a group of operations that cannot be described by a continuum of numbers. For example: The rotations around a fixed axis can be described by a *continuum* of angles running from 0° to 360°; however, the rotations that leave a square looking the same are the *discrete* set of 90°, 180°, 270° and 360° (equivalent to 0°).

down a flavor of quark that, together with the *up* quark, makes up all of nuclear matter.

effective mass for a collection of (usually two or three) particles: the mass that a particle must have to have decayed into that collection of particles.

effective theory a theory whose range of validity does not extend above a certain energy, so that it is only an *approximation* to the fundamental theory, and need not be renormalizable.

electric field a field of disturbances that interacts with *charged* particles in the region of the field. It is usually caused by charged particles but can exist independently as part of an electromagnetic field.

electromagnetic field coupled electric and magnetic fields that sustain one another's existence. It is also used as the generic term to describe any combination of electric and magnetic fields.

electromagnet a device that produces a magnetic field from currents flowing through coils of conducting wire.

electromagnetic interaction the basic interaction, mediated by photons, responsible for electrical, magnetic and optical phenomena.

electron a negatively charged fundamental weakly interacting particle (lepton) of nature found in every (neutral) atom. The flow of electrons in wires is commonly known as electricity. It has spin $\frac{1}{2}$, so it is a fermion, and has lepton number $+1$.

electrostatic involving only charges at rest.

electroweak theory the theory that unifies electromagnetic theory and the weak-interaction behavior of particles. It is the gauge theory $SU(2)_L X U(1)$.

energy a property of objects, related to their motion, locations and masses. The total energy of the universe is believed to be conserved — energy can only be transferred from one object or system to another. Work is the transfer (by the exertion of a force) of energy from one object to another. There are many types of energy: motional — called kinetic, positional — called potential, mass — called rest energy, etc.

eta (η) or **eta zero (η°)**, also called **eta naught** a meson that is in the same FLAVOR $SU(3)$ octet as the pion and kaon.

eV electron volts. It is the amount of kinetic energy an electron gains as it moves through 1 volt.

exclusion principle a law of nature that prevents two identical fermions from being in the same quantum state.

Fermilab (Fermi National Accelerator Laboratory or FNAL) The high-energy particle accelerator national laboratory located in Batavia Illinois (near Chicago).

fermion a particle whose spin is an odd half-integer (times \hbar), e.g., $1/2, 3/2, 5/2$.

field conglomeration of disturbances, one at each point in a region of space, which would exert forces on particles at those locations; the set of those disturbances over a continuous range of positions.

field lines representative lines drawn to show the directions and relative strengths of the disturbances of a field.

flavors of strongly interacting particles: properties that determine particle type, e.g., isospin, hypercharge, strangeness, and charm. They originate from the quark flavors (types) that the particle is made of. This word is also used to characterize the six different types of quarks. of leptons: sometimes used to characterize the six different types of leptons (which do not interact strongly) as well.

flavor space a mathematically-defined space in which locations correspond to values of flavor numbers, rather than distances from an origin. See *internal space*.

FLAVOR $SU(3)$ The *simple* group containing an entwining of the two flavor groups: $SU(2)$ of isotopic spin and $U(1)$ of flavor-hypercharge. It is a broken symmetry due to the different masses of the *up, down,* and *strange* quarks.

FLAVOR $SU(2) \times U(1)$ a mere joining together as *independent* groups of the two flavor groups: $SU(2)$ of isotopic spin and $U(1)$ of flavor-hypercharge.

flux tube in quantum chromodynamics: color-field lines, running from one particle to another, that are confined to a thin region of space which consequently has a tube-like shape.

focusing in particle accelerators: methods used to keep the particles of the beam from deviating so far from the ideal beam path that they hit the walls of the beam pipe.

force a push or pull on an object.

frequency the number of oscillations per second. It is the inverse of the period of oscillation.

fundamental theory a theory that has no energy limitation to its validity, so that it must be renormalizable.

gauge bosons quantum particles of an interaction; they carry the interaction from one particle to another. They all have spin 1, except for the graviton which has spin 2. For unbroken gauge theories, the gauge particles are massless. Massless examples are: the photon (of the electromagnetic interaction) and the gluons (of the strong interaction — QCD). The Ws and Z are massive due to symmetry breaking.

gauge symmetry see gauge theory.

gauge theory a theory that is invariant to *local* transformations (i.e., that may be arbitrarily different at different points in space) of an internal symmetry group, where the gauge particles absorb the resulting "stress."

gaugino particles that are the spin-1/2 supersymmetric partners of the (spin-1) gauge particles in supersymmetric gauge theories. See also *photino, Wino, Zino.*

generation the three isodoublets of quarks are considered three separate generations, the first generation corresponding to the lightest doublet, etc. The same is true for the leptons.

generator a simple operator (usually representing a physical quantity) that determines the effect of a continuous set of physical operations, e.g., momentum is the generator of translations and angular momentum is the generator of rotations.

GeV Giga-electron volts, 10^9 eV.

glueballs a bound state of two gluons. Also called *gluonium.*

gluons gauge bosons of the strong interaction; they carry the strong interaction from one particle to another. There are eight different types of gluons.

gluonium a bound state of two gluons. Also called *glueballs.*

grand unified theories theories which contain the Standard Model, but go beyond it by *unifying* the strong and electroweak forces.

gravitational interaction the interaction that causes every mass to pull on every other mass; the cause of gravity.

gravitino the spin-3/2 supersymmetric partner of the (spin-2) graviton in supergravity theories.

graviton gauge boson of the gravitational interaction. The only "known" gauge boson with spin 2. See *quantum particles of gravity.*

group a set of operations such that (1) when two operations are performed in succession, the result is equivalent to an operation of the group, e.g., a rotation about an axis by 30° followed by a rotation about the same axis by 20° is equivalent to a rotation by 50°; (2) doing nothing to the system being operated on is included in the set (i.e., a rotation by 0°); (3) for every operation in the set there is an opposite operation in the set; (4) a technical mathematical requirement called associativity, which does not concern us here.

GUTs acronym for Grand Unified Theories.

h Planck's constant.

hadron strongly interacting particles; made of quarks and/or gluons.

hadronic phase the phase of matter in which there are no free quarks or gluons; they are all bound into (colorless) hadrons or glueballs. Our present-day universe is in the hadronic phase of matter.

hadronization a multitude of field-theoretic processes (involving virtual particles) that replace each quark, antiquark and gluon (which are particles of color) that are knocked out of hadrons by a jet of (colorless) hadrons. Only these colorless jets can emerge. See *confinement.*

half-life the length of time during which the probability of decay of an unstable particle is 1/2. For a large number of such particles, about one half of them will decay during a time interval of one half-life.

handedness the circumstance in which right-handed particles have different properties from left-handed ones. Also, used to refer to right-handedness or left-handedness.

Higgs field the field associated with the Higgs mechanism; it should permeate all of space, and is thought to be the source of the masses of the fundamental particles. See *Higgs particle and Higgs mechanism.*

Higgs mechanism a mechanism that causes the *breaking* of a symmetry of the interactions *by the physical states* of the system, leaving the interactions symmetric. See *Higgs field.*

Higgs particle The quantum particle of the Higgs field associated with the Higgs mechanism. See *Higgs field* and *Higgs mechanism.*

hypercharge (Y) two *times* the amount that the electric charges of the particles of an isotopic spin or isospin (depending on the context) multiplet are shifted from their corresponding I_z values. Alternatively, two *times* the amount that the average charge of a multiplet is shifted from 0. The hypercharge originally associated with the strong interaction (referred to as "flavor-hypercharge" in this book) has been generalized so that it applies to both quarks and leptons.

indeterminacy the fundamental lack of precise predictability of the evolution of a system. It is a property of probabilistic theories of nature.

interaction a general term that includes force, particle creation and particle annihilation.

internal space a mathematically-defined space (of our imaginations) in which locations correspond to values of particle-type quantum numbers, rather than distances from an origin. Despite the presence of the word "internal," these are *not* physical spaces inside *anything*; instead "internal" is being used to indicate that it corresponds to particle properties rather than spacetime.

internal symmetries symmetries comprised of operations that change particle-types; these are operations that are done mentally, within our imaginations (and mathematically in the corresponding internal space). See also *internal space* and *flavor space*.

inversion reversing all three directions of space; equivalent to a reflection plus a rotation. see *reflection (inversion) symmetry*.

ion the remaining object after one or more electrons have been removed from an atom. It consequently has less electrons than protons, so that it is positively charged.

ionization the process of removing one or more electrons from an atom.

isospin a generalization of isotopic spin that applies to all the quarks and the leptons as well. The $SU(2)$ gauge theory associated with isospin is part of the electroweak theory.

isotope nuclei that have the same number of protons but a different number of neutrons are *isotopes* of one another.

isotopic doublet an isotopic spin $1/2$ multiplet; it contains two particles, one with isotopic spin up ($I_z = +1/2$) and one with isotopic spin down ($I_z = -1/2$). The familiar example of an isotopic spin $1/2$ multiplet is the nucleon doublet comprised of the proton, with isotopic spin up, and the neutron, with isotopic spin down.

isotopic spin a flavor property of particles that is mathematically analogous to spin, but exists in an imagined space. Its allowed values are 0, $1/2$, 1, etc., in steps of $1/2$. The isotopic spin properties of a set of particles originate from the *up* and *down* quarks among their constituents. It has *nothing* to do with isotopes of elements. See *isotopic doublet*, *isotopic triplet*, and *isotopic quartet*.

isotopic triplet an isotopic spin 1 multiplet; it contains three particles, one with $I_z = +1$, one with $I_z = 0$ and one with $I_z = -1$.

I_z the component of isotopic spin or of its generalization isospin (depending on the context) in the "z-direction" in the space associated with the isotopic spin or the isospin. The charge of a particle is the sum of its I_z-value and one half of its hypercharge ($1/2Y$).

jet in high-energy experiments: a stream of high-energy particles emerging from an interaction, that is presumed to have arisen from a scattered quark or gluon.

Kamiokande a huge neutrino detector located in a mine in Kamioka, Japan.

kaon (K) or **K meson** a meson produced in elementary-particle collisions. It is in the same FLAVOR SU(3) octet as the pion.

keV kilo-electron volts or 10^3 eV.

lambda (Λ) or **lambda zero (Λ⁰), also called lambda naught** a heavy baryon that is in the same SU(3) octet as the nucleon (proton and neutron).

left-handed for particles: a particle that behaves as if it is spinning around its momentum in the direction that a left-handed screw is turned to tighten it.

lepton number (*L*) the total number of leptons *minus* the total number of antileptons.

leptonic decay a decay of a particle into leptons.

leptons fundamental matter particles, that do not have strong interactions; there are six types altogether: electron, muon, tau and their corresponding neutrinos. They all have spin 1/2, so they are fermions. The left-handed ones all have weak interactions.

LHC the Large Hadron Collider located at CERN in Geneva, Switzerland.

lifetime for unstable particles; the average time that such particles exist before decaying.

linac an acronym for *linear accelerator*.

linear accelerator an accelerator in which the beam particles make one traversal of a straight path.

linear collider a modified linear accelerator which produces two beams that are bent so that they collide head-on with one another.

local transformations "internal" transformations that are *different* at different space-time points. When a gauge field is present to absorb the resulting "stress", we have a gauge theory.

luminosity for colliders; a quantity that represents the effective combined "brightness" of the beams, such that the luminosity *times* the cross section (for the process under study) gives the number of events that will occur per second.

mass related to the weight of an object, but does not depend on the strength of gravity. For macroscopic objects it is a measure of the amount of material; for individual particles as well as large objects, it is a measure of their rest energies.

mechanistic philosophically: an adjective describing a world that follows fixed laws of nature and consequently runs like a machine.

meson subatomic, strongly interacting particles that have integral spin (bosons). They are composed of one quark and one antiquark.

MeV million electron volts or 10^6 eV.

micro- prefix meaning one millionth.

mirror image the description of what would be observed if an event were observed in a mirror.

momentum a quantity that is a measure of the motion of an object, and the impulsive force it can exert. It is the product of the particle's mass and its velocity.

Monte Carlo program a computer program written to simulate a scattering or decay experiment, in which the random locations of the beam particles and target particles are mimicked by the use of random number generators. It is named after that famous gambling mecca where it is believed that the numbers that come up are indeed random.

muon (μ) a fundamental weakly interacting (lepton) particle of nature; much like an electron (also negatively charged with spin 1/2), but more than 200 times as massive.

neutral having zero charge

neutrino an uncharged fundamental particle of nature that does not experience strong interactions (i.e., a lepton); there are three types: electron-neutrino, muon-neutrino, tau-neutrino. In the Standard Model, they are massless particles, but recent experiments indicate that they may have tiny masses.

neutrino oscillations the changing of one type of neutrino into other types and back again, as time goes on.

neutron an electrically neutral constituent of the atomic nucleus; the neutron is composed of three quarks. (It is the isotopic-spin down member of the isotopic doublet known as the *nucleon*.)

nonleptonic decays the decays of hadrons to other hadrons.

nucleon a proton or a neutron. (Together, they are classified as *the nucleon* — an isotopic-spin 1/2 doublet.)

omega minus (Ω^-) or **omega (Ω)** the most massive baryon in the lightest SU(3) baryon decuplet. It is composed of three *strange* quarks. Its existence was predicted by broken FLAVOR SU(3), and that discovery displayed the legitimacy of that model.

Parity (P) the operation of changing the coordinate system to one inverted in all three dimensions, equivalent to reflecting in three orthogonal mirrors. The behavior of an object's wavefunction under this operation.

peak the highest part of a wave form.

period the time it takes for one complete oscillation to occur. It is the inverse of the frequency.

perturbative an adjective that characterizes a solution (here for a quantum field theory) that is a succession of smaller and smaller terms; each successive term improves the accuracy of the result. The use of Feynman diagrams, with their inherent virtual particles, to generate a solution or prediction is described as a perturbative solution.

phase of a wave: any single specific part of the wave's oscillation. of matter: qualitative molecular arrangements that correspond to solid, liquid, gas, and others less well-known. Also, see *hadronic phase*.

phase transition the changing of the phase of matter, such as melting (a solid-to-liquid phase transition).

photino a particle that is the spin-1/2 supersymmetric partner of the (spin-1) photon in supersymmetric gauge theories. See also *gaugino, Wino, Zino*.

photon a quantum particle of the electromagnetic interaction (including light); it is the gauge boson of the electromagnetic interaction that carries the interaction from one charged particle to another. In some situations it can produce an electron and positron in its stead.

pion (π) the least massive meson, composed of a quark and an antiquark; it is commonly produced in particle collisions.

Planck's constant (h) a physical constant discovered by M. Planck, which determines the sizes of the steps in the allowed values of quantized physical quantities.

positron the antiparticle of the electron; it is positively charged with lepton number -1, but otherwise identical to an electron.

probabilistic philosophically: an adjective describing a world in which the *present* determines only the relative probabilities of several possible *futures*.

proportional counter a gas-filled particle detector chamber consisting of a wire and a metal plate or tube with a large voltage between them. As charged particles pass through the chamber they ionize atoms of the gas. The voltage is large enough so that those ions and the freed electrons then ionize more atoms, etc., leading to an avalanche of ions and electrons. Multiwire proportional counters have many wires and plates.

proton a positively charged strongly-interacting particle. Protons are usually found inside atomic nuclei. A proton *is* the nucleus of a hydrogen atom. The proton is composed of three quarks. It is the isotopic-spin up member of the isotopic doublet known as the *nucleon*, and is the lightest baryon (strongly-interacting fermion).

quantized to have allowed values that are discrete, rather than continuous.

quantum an adjective used to describe quantities that come only in discrete amounts.

quantum chromodynamics (QCD) the quantum theory that describes the strong interaction. It is a gauge theory based on (color) SU(3) with gluons as gauge particles.

quantum conjugate a relationship between two observables, like position and momentum, that obey an uncertainty relation. The narrower the region of uncertainty of one of them, the wider is that of the other.

quantum electrodynamics (QED) the quantum theory that describes the electromagnetic field as a superposition of photons, and the interaction between charged particles in terms of their exchange of photons. It is a U(1) gauge theory with the photon as the gauge particle.

quantum field an apparent field of disturbances, that behaves like a superposition of virtual quantum *particles*.

quantum mechanics the probabilistic quantum theory that, with its extension to quantum field theories, succeeds (so far) in explaining our observations and experiments.

quantum particle a particle that has classical wave properties *as well as* classical particle properties, and is described by quantum mechanics. All of the particles of nature are quantum particles.

quantum particles of gravity gravitons; the particles whose exchange between particles possessing energy (including their rest energies, i.e., masses) causes the force of gravity.

quantum theory a theory of nature that has quantized steps for the allowed values of many properties of particles.

quark-flow diagrams diagrams that keep track of the incoming and outgoing quarks of a reaction, using the simplest choice of internal quark and gluon lines solely for that purpose.

quark-gluon plasma a state of matter in which quarks and gluons exist separately, not bound together in hadrons. It is believed to have been the state of the universe shortly after the Big Bang.

quarks the fundamental particles of which all strongly interacting particles (hadrons) are composed. There are six different flavors of quarks and each flavor comes in three colors.

reflection group the set of operations: reflect (in a mirror), do not reflect (which is equivalent to a double reflection).

reflection (or inversion) symmetry an interaction that appears unchanged when looked at in a reflected (or inverted) coordinate system has reflection (or inversion) symmetry. When a reflected (or inverted) set of events is indistinguishable from some unreflected set of *possible* observations, so that we cannot tell that that data is the result of reflection (or inversion) of *actual* data, we have reflection (or inversion) symmetry.

relativistic mass the energy of a particle divided by the square of the speed of light (E/c^2).

renormalizable a theory is renormalizable if it can be successfully renormalized. See *renormalization.*

renormalization an adjustment performed on the calculations in quantum field theories that lumps all of the interactional modifications of the charge and mass (and perhaps a *finite* set of other constants) of each particle together with their initially symbolic values (represented merely by algebraic letters). All together, these mathematical lumps are taken to be the *observed* values of the respective quantities.

renormalize here, to apply a renormalization procedure to a quantum field theory.

resonance a response to an incident disturbance or wave that takes a large amount of energy from it. In atomic and nuclear processes, that energy causes transitions to other states or enhances scattering cross sections. In particle physics, the word "resonance" is used to designate very short-lived particles that only show their existences as peaks in plots of the number of events (for a spread of effective-mass values of their decay products) or in appropriate cross sections.

rest mass or **(particle) mass** the intrinsic mass of a particle; it is often used to identify the particle. A particle's rest mass corresponds to its rest energy divided by the square of the speed of light or to its relativistic mass when it is at rest.

rf stands for *radio frequency* and refers to any electromagnetic waves or voltages whose frequencies are in the same range as those used to transmit radio and television.

RHIC acronym for Relativistic Heavy Ion Collider. It is a collider, recently built at Brookhaven National Laboratory in Long Island, NY, that accelerates heavy ions to high speeds and studies their collisions (with one another) within its detectors.

rho (ρ) a very short-lived (spin-1) meson, sometimes referred to as a resonance. It usually decays into two pions. See *resonance.*

right-handed for particles; a particle that behaves as if it is spinning around its momentum in the direction that a right-handed screw is turned to tighten it.

rotational symmetry the property of systems or interactions of appearing the same, when viewed in a rotated orientation, as they did before rotation.

scalar an object unaffected by transformations of a symmetry group. Usually, when the group is not specified, the group is understood to be the group of rotations. Also, a spin 0 particle.

scintillator a material that gives off visible light when a charged particle or photon (ionizing particles) moves through.

scintillation counter a particle detector that detects light produced in scintillator material to determine energy. See *calorimeter.*

selectron the (spin-0) partner of the electron in supersymmetric theories.

sigma (Σ) a heavy baryon produced in nuclear collisions. It has three possible charges, so that this is the generic name for the Σ^+, Σ^0 and Σ^-. They are members of the lightest Flavor SU(3) octet.

silicon-strip detector a particle detector that utilizes a small silicon strip (like that used to make transistors).

simple group here, a group of operations whose generators are fully intertwined. No "×" appears in the name. More technically, it is a group that does not have a (non-trivial) invariant subgroup.

singlet (1) the name used for a multiplet of any group that contains only one particle; that particle is consequently invariant to the operations of the group.

SLAC acronym for Stanford Linear Accelerator Center, located in Stanford, California.

SLC acronym for Stanford Linear Collider.

sneutrino in supersymmetric theories; the (spin-0) partner of a neutrino.

SNO acronym for Sudbury Neutrino Observatory, a huge neutrino detector located in a mine in Sudbury, Ontario, Canada

SO(3) the mathematical name of the rotation group in three dimensions.

SO(10) the mathematical name of the group used in one of the proposed grand unified theories.

solar-neutrino problem the paucity of electron-neutrinos coming from the sun, as compared to the prediction from the standard solar model. Neutrino oscillations have been proposed as the explanation.

solar wind the stream of charged particles (mostly protons and electrons) sweeping out of the sun towards all of the planets. Sometimes the words "cosmic rays" are used loosely to include the particles in the solar wind with all of the others that strike the Earth's atmosphere.

spaghetti calorimeter a metal rod with scintillator material embedded lengthwise, whose end looks like the open end of a box of spaghetti, used to measure the energies of charged particles.

spark chamber a gas-filled detector containing wires at different voltages in which visible sparks appear along the trail of a charged particle that traveled through the detector.

spectator quarks in high-energy hadron collisions, those quarks that do not undergo the basic "hard" scattering.

spin intrinsic angular momentum possessed by particles. It is quantized in units of $h/2\pi$ represented by "\hbar" (called "h-bar"), with possible values of 0, 1/2, 1, 3/2, 2, ..., etc., (times \hbar).

squark in supersymmetric theories; the (spin-0) partner of a quark.

SSC acronym for Superconducting SuperCollider, a giant cyclic collider that was under construction in Texas in the early 1990s when it was canceled by Congress.

stable for particles in general: one that does not decay (therefore, does not emit radiation) is called stable. In elementary particle physics, those that live long enough to produce measurable tracks are referred to as stable particles.

storage ring a ring-shaped or oval cyclic particle collider, used to "store" more and more bunches of particles to increase luminosity.

strange the name of the type (flavor) of quark responsible for the strangeness properties of the particles that contain it.

strangeness (S) a chargelike flavor number that is conserved in the strong interactions. It is part of the flavor-hypercharge.

streamer chamber very much like a spark chamber, but producing many very short sparks.

strong interaction the strongest interaction between particles, responsible for nuclear forces. It is carried by gluons from one quark (or antiquark) to another, or between hadrons (particles composed of quarks and/or antiquarks).

SU(2) the mathematical name of the continuous group of operations that corresponds to three intertwined generators. It is an extension of the three-dimensional rotation group to include particles of spin 1/2, 3/2, etc. — the fermions. Both isotopic spin and its generalization isospin are also SU(2) groups.

SU(2), FLAVOR the mathematical name of the group of isotopic spin rotations.

SU(2) × U(1), FLAVOR a group that consists of the group of isotopic spin (SU(2)) independently combined (not intertwined) with the group U(1) of flavor-hypercharge rotations.

SU$_L$(2) × U(1) or **SU$_L$(2) × U$_Y$(1)** the gauge group that underlies the electroweak theory — part of the Standard Model.

SU(3) antitriplet ($\bar{3}$) a multiplet containing three particles that are the antiparticles of those in the triplet. For FLAVOR SU(3) the antitriplet contains three flavors of anti-quarks, viz., *anti-up*, *anti-down*, and *anti-strange*; for COLOR SU(3) the antitriplet contains three colors called *anti-red*, *anti-green*, and *anti-blue*.

SU(3), COLOR see Color SU(3).

SU(3), FLAVOR see Flavor SU(3).

SU(3) decuplet (10) an SU(3) multiplet containing ten particles. Some sets of baryons form decuplets of FLAVOR SU(3).

SU(3) octet (8) a multiplet of SU(3) containing eight particles. Most hadrons belong to octets of FLAVOR SU(3). For COLOR SU(3), the (8) gluons also form an octet.

SU(3) triplet (3) a multiplet containing three particles, For FLAVOR SU(3) the triplet contains three flavors of quarks, viz., *up*, *down*, and *strange*; for COLOR SU(3) the triplet contains three colors called *red*, *green*, and *blue*.

SU(3) × SU(2) × U(1) or **SU(3) × SU$_L$(2) × U$_Y$(1)** or **SU$_c$(3) × SU$_L$(2) × U(1)**, etc., also called **3-2-1** the gauge group of the Standard Model, composed of the SU(3) of color, for the strong interactions, and the SU(2) × U(1) electroweak theory.

SU(5) the mathematical name of the group used in the simplest grand unified theory.

SUGRA acronym for SUperGRAvity; refers to *local* supersymmetric theories, which all contain gravity in them as an automatic consequence of their *local* symmetry.

superconducting magnet an electromagnet (coils of wire that produce a magnetic field when current flows through) made of superconducting material. It must be kept at temperatures very close to absolute zero to remain superconducting.

supergravity see *SUGRA*.

Super-Kamiokande souped-up name for the upgraded Kamiokande neutrino detector. See *Kamiokande*.

superpartner in supersymmetric theories; the particle associated with a particular particle by supersymmetry operations, whose spin differs by $1/2$ from that of the original particle. No superpartner of any particle we know has been found (as of this writing).

superstrings in superstring theories: the fundamental entities of nature. They *look* like *particles* at all energies or temperatures below that of the very very earliest instant after the Big Bang.

supersymmetry a symmetry whose transformations mix bosons and fermions.

SUSY acronym for SUperSYmetric theory.

symmetry an operation or set of operations that can be performed on a system or on a theory leaving it unchanged.

synchrotron a particle accelerator with an oval beam path that the beam particles traverse over and over again. The acceleration is provided in small increments during each traversal by electromagnetic waves, and magnets are used to keep the particles on path.

synchrotron radiation electromagnetic radiation emitted by a charged particle when it is moving in a circular or oval path.

tau (τ) a fundamental weakly interacting particle of nature (lepton); much like an electron and muon (also negatively charged with spin $1/2$), but very much more massive than either.

three-dimensional rotation group the set of all of the rotations around a point. The mathematical name of the group is SO(3).

time-projection chamber a particle detector that utilizes timing to determine the location in one of the three directions.

Time reversal (T) when looking at a reaction or state: considering what the situation would be for a reversed direction of the flow of time — running the "film" backwards.

top the flavor of the heaviest quark, also known as *truth*.

tower calorimeter a calorimeter constructed of alternating layers of metal and scintillator.

translate (spatial) in mathematics: to move an object, system, or frame of reference from one place to another without changing the orientation, i.e., without rotating it.

trough the lowest part of a wave form.

truth the flavor of the heaviest quark, also known as *top*.

two-dimensional rotation group the set of rotations around a fixed direction. It leaves objects that are in any plane perpendicular to the fixed direction in that same plane, hence the name. The mathematical name of the group is U(1).

U(1) the mathematical name of the continuous group of operations that corresponds to a single generator. The rotations about a fixed direction in space (two dimensional rotations) form a U(1) group of rotations in space; the group generated by hypercharge and the group generated by electric charge also form U(1) groups (in their respective conceptual spaces).

uncertainty principle in quantum mechanics: a principle which says that certain properties of particles cannot be determined simultaneously; the more precise the measurement of one, the more spread are the possible values of the other. See *uncertainty relation*.

uncertainty relation from quantum mechanics: the relationship that states that the product of the spreads of the measured values (from many identical systems) of two different quantities must exceed $1/2\ \hbar$ — the more definite the value of one quantity, the more uncertain (spread) is the value of the other. It only occurs when the result of successive measurements of each of those two quantities differs from the result of making those measurements in reversed order. See *uncertainty principle* and *quantum conjugate*.

universe all of space.

unrenormalizable a theory is unrenormalizable if it cannot be successfully renormalized (by use of a finite set of observable numbers).

unstable here, an adjective referring to a type of particle or nucleus that changes ("decays") into other types of particles or nuclei.

up a flavor of quark that, together with the *down* quark, makes up all of nuclear matter.

vacuum this word is used in two different senses here: 1. "The ideal vacuum" is a hypothetical state of the world that is completely empty, with no particles or fields. 2. When particles are present, the parts of space that have no particles or flux-lines of forces are referred to as *vacuum* regions. In quantum field theories, one must perform renormalizations of the calculations, which always leave residual effects that

appear to be due to the creation and annihilation of virtual particles in the vacuum (regions). These effects are commonly referred to as *vacuum polarization effects*.

vacuum polarization effects See *vacuum*.

vector meson a meson that has spin 1.

veto in large particle detectors: utilization of observations made by a part of the complex detector to prevent the turn-on of the *whole* detector or the recording of the event.

virtual *Virtual* particles appear in perturbative treatments of quantum field theories. They are not detectable as particles, but rather contribute to the interaction between the real particles that are present. In vacuum regions, virtual pairs of particles are created and annihilated randomly, affecting the properties of the actual particles present.

W a massive charged gauge boson of the weak interaction that carries the weak interaction between particles. It mediates beta (β) decay.

wave classical: an oscillatory disturbance that is spread over a continuous region of space.

wavefunction in quantum mechanics: a mathematical object whose (absolute) square yields the probability distribution of the position of a particle or particles.

wavelength for *simple* (sinusoidal) waves: the distance between two successive peaks or the distance between any part of the wave and the successive repetition of that part.

weak decays the changing of a particle into a less massive one plus a pair of leptons or quarks, due to the weak interaction.

weak interaction the interaction among particles that is responsible for most radioactivity, including particle decays. The Ws and the Z carry the interaction between particles.

weight the pull of gravity on an object. An object's weight depends on the strength of gravity at its location.

white the designation of a particle or set of particles that are unchanged by COLOR-SU(3) transformations, i.e. a color singlet. Also referred to as *colorless*.

width of data in particle experiments or of a peak in effective-mass plots for a (small) set of particles; it is inversely related to the finite (usually extremely short) lifetime of a particle (or resonance) that decayed into those observed particles.

Wino (pronounced "weeno") a particle that is the spin-1/2 supersymmetric partner of the (spin-1) W in supersymmetric gauge theories. See also *gaugino, photino, Zino*.

"×" a designation used between the names of groups to indicate that their respective operations are independent of one another.

Y the symbol for hypercharge.

Z a neutral massive gauge boson of the weak interaction that carries the interaction between particles.

Zino a particle that is the spin-1/2 supersymmetric partner of the (spin-1) Z in supersymmetric gauge theories. See also *gaugino, Wino, photino*.

Index